Integrated SCIENCE 1

George Bethell
David Coppock

OXFORD

Integrated SCIENCE 1 141

Oxford University Press, Walton Street, Oxford OX2 6DP

Oxford New York Toronto
Delhi Bombay Calcutta Madras Karachi
Kuala Lumpur Singapore Hong Kong Tokyo
Nairobi Dar es Salaam Cape Town
Melbourne Auckland Madrid

and associated companies in
Berlin Ibadan

Oxford is a trade mark of Oxford University Press

© George Bethell, David Coppock

ISBN 0 19 914268 8

Reprinted (with corrections) 1992, 1993

Typeset in Plantin Light by
Tradespools Ltd, Frome, Somerset
Printed in Hong Kong

Acknowledgements

The publishers wish to thank the following for permission to reproduce photographs:

Aga-Rayburn: p 74 (left); **Ace Photo:** p 162 (bottom left); **Allsport/Steve Powell:** pp 8 (top left), /**Tony Duffy** 54 (right); **Heather Angel:** pp 115 (right), 130 (top right), 130 (right), 227 (top right), 231 (left), 241 (top right); **Aspect Picture Library:** pp 28 (bottom right), 64 (centre right), 131 (centre right), 207; **BBC Hulton Picture Library:** pp 78, 97(top right), 109 (right), 111 (top right), 239 (right, top right); **Biophoto Associates:** pp 36 (right, top right); **Brenard Ltd:** p 74 (centre right); **British Drag Racing Association:** p 57 (left); **British Nuclear Fuels:** pp 107 (right), 131 (right); **British Petroleum:** p 74 (centre); **Calor Gas Ltd:** p 74 (centre left); **Camera Press:** p 179 (top right), 180 (right), **J Allan Cash:** pp 68 (top right), 69 (left, top right), 75 (right), 118 (centre), 162 (left), 185 (top right), 201, 237, 238 (bottom right), 241 (right); **Casio Electronics Ltd:** 91; **Central Electricity Generating Board:** p 143; **Civil Aviation Authority:** p 86; **John Cleare:** p 165 (right); **Bruce Coleman Ltd:** pp 66 (bottom right), 70 (left), 85 (centre), 90 (left), 114 (bottom right), 115 (bottom right), 130 (left), 131 (centre), 163 (top right), 187 (right), 221 (bottom), 238 (top right), 240; **Colorsport:** pp 8 (centre, top right, centre right), 38 (top right), 48 (top), 56 (right), 57 (right), 177 (right, centre right); **Mary Evans Picture Library:** pp 190, 132; **Ford Motor Company:** p 47 (right); **GeoScience Features:** p 233 (top right); **Gibbs Oral Services:** p 13 (bottom right); **Sally & Richard Greenhill:** pp 51 (right), 66 (top right), 165 (centre right), 184; **Susan Griggs Agency:** pp 66 (left), 155 (centre), 221 (top); **Philip Harris:** pp 102, 216; **Health Education Council:** p 20 (top right); **Hotpoint:** p 172; **ICI:** p 123; **Intercity:** p 74 (right); **ITN:** p 205; **London Fire Brigade:** p 85 (left, centre); **Laurie Morton:** pp 240 (bottom right), 245; **NASA:** pp 43 (centre right), 210; **National Medical Slide Bank:** p 23 (bottom left); **Natural History Museum/Martin Poulsford:** pp 61 (right), 62 (centre); **Oxford Scientific Films:** pp 21 (top left), /**T Middleton** 114 (right), /**Press Tige** 116 (right), /**B Watts** 124, 226 (top right, left), /**C Gardener** 226 (bottom right), /**B Watts** 227 (centre right), 230 (top right), /**J Cooke** 230 (bottom right), 239 (bottom right); **Phillips:** p 154 (top right); **Picturepoint:** p 64 (bottom right); **Press Association:** p 38 (left); **Mike Roberts:** pp 38 (top left), 44, 245; **Science Photo Library/Michael Abbey:** pp 21 (top left), /23 (bottom right), /**Petit Format** 31 (all), /**Carolyn Jones** 32 (bottom right), /**National Cancer Institute** 35 (top left), /**Dr T Brain** 35 (top), /**Dr J Lorre** 38 (right), /**Dr J Burgess** 180 (top right), 188 (top), 212 (left), /**Doug Allan** 212 (right), /**Angela Murray** 212 (left), 214, 220, /**M Abbey** 225 (top right), /**E Grave** 225 (left), /**J Walsh** 225 (left), /**E Grave** 226 (top right), 230 (right); **Shell:** pp 80 (top right), 81 (top right); **Barry Stone:** p 213; **Tass:** p 108; **The Sunday Times/Ian Yeomans;** p 23 (top right); **United Kingdom Atomic Energy Authority:** 97 (right), 98, 103 (centre right), 130 (centre); **Wind Energy Group Ltd:** p 90; **World Wide Fund For Nature:** p 132; **Zefa Photographic Library:** pp 20 (bottom right), /**J Pfaff** 66 (right), 68 (bottom right), 89 (right), /**Schlodien** 114 (centre right), /**W Harstrick** 115 (top right), /**E & P Bauer** 118 (left), /**Bob Croxford** 118 (right), /**Hinz** 121 (top right), /**Dr Baer** 121 (right), 130 (centre, right), /**R Smith** 130 (centre), 131 (left), /**Hackenberg** 131 (left), /**Deuter** 154 (right), /**G Herman** 155 (right), 174 (left), 240, /**W Hamilton** 240.

Additional photography by **Peter Gould**. Illustrations by **Brian Beckett, Elitta Fell, David La Grange, Nick Hawken, Jones Sewell, Alan Rowe, Julie Tolliday, Galina Zolfaghari**.

Introduction

This book has been written for students like you, studying science at (senior) secondary level. It contains 12 chapters covering many important topics, particularly those that affect us in our everyday lives. The topics are usually presented over two or four pages and there are lots of diagrams and photographs to help you. There are also questions to check that you have understood the main ideas. Where the topics overlap cross-references are given to guide you.

Of course, science is more than just reading books, however good they are, and your school's programme of practical work will help you develop experimental skills. To support this you will find 'activity items' throughout the book. These are things that you can do at school under the guidance of a teacher or at home on your own. Some of the activities suggest experimental work, others ask you to carry out surveys or collect data from sources such as newspapers and magazines. They are all there to help you develop a scientific attitude.

We obviously hope that this book will help you to pass your examinations but we also hope that it will increase your interest in science and its applications. It does contain many examples of how modern technology has improved our lives but it also considers the harmful effects of the way in which we use science. It tries to give information and a balanced view on some of the most important issues of the day; personal health, environmental pollution, nuclear power and the energy crisis, and others. Using this information you will be able to make up your own mind and take part in discussions on these matters as a well-informed person.

Above all, we hope that you will enjoy using this book and that you will finish it feeling more confident that you can use scientific ideas and methods to understand the way our world works.

George Bethell   **1990**
David Coppock

Contents

What gives the body its shape?

How do we move?

How do we get information about the world?

How do we stay healthy?

How do we reproduce?

These athletes work hard to keep their bodies healthy. There are lots of differences between them. They are different sizes and shapes. They have different skin colours. However, their bodies work in exactly the same way. They need the same type of foods, they move in the same way and they can catch the same kind of diseases.

It is important to study how the body works. We can use the information to prevent injury at work or during sport. We can also reduce the risk of disease. By keeping our bodies healthy we can lead much more enjoyable lives.

Remember, you only get one body . . . look after it!

Activities

Answer these questions about your body:

1   How tall are you? (Give your answer in metres and centimetres.)

2   How much do you weigh? (Give your answer in kilograms.)

3   How would you describe yourself?
     tall / medium / short
     overweight / about right / skinny
     athletic / average / not athletic

4   How would you describe your health?
     very healthy / about average / unhealthy

5   How many times have you been ill in the last three months?
     0–1 / 2–3 / 4 or more

6   **a)** How many teeth do you have?
     b) How many have been filled?

7   Do you wear glasses or contact lenses?

8   How much exercise do you take?
     lots / about average / not very much

The human machine

Human beings, like most large animals, have a **skeleton** inside their bodies. A skeleton is a system of bones and other supporting material and has three important functions:

- it gives support to the rest of the body – like the framework of a building. This gives the body its shape.
- it gives protection to important and delicate organs of the body, for example, the skull protects the brain.
- it provides an anchorage for muscles. Muscles fixed to the skeleton can operate **joints**. This makes parts of the skeleton move.

From the diagram of the skeleton, you can see how the skull gives the head its shape. The brain is protected by the bones of the skull. The eyes are protected by bony sockets. Muscles attached to the jaw allow us to eat and speak.

Bones

In the human skeleton there are over two hundred bones. Some are long, some short, some round, some flat, but all bones have the same basic structure.

As a baby develops inside its mother's womb some cells form a tough, flexible substance called **cartilage** (gristle). (You can feel cartilage in your ears and at the end of your nose.) Slowly much of the cartilage changes to bone. Bone is very hard and strong. It has to stand up to large forces. Large bones have a hollow shaft inside them which makes them lighter. This makes movement easier. The bone **marrow** within the shafts produces blood cells.

Bones have living parts and non-living parts. The living part makes the bones slightly flexible. This lets them absorb sudden shocks. The bones of old people have less living material and so they become **brittle** and break easily.

The non-living part of a bone makes it stiff (rigid) and gives it strength. This can be shown by putting a bone in a beaker of dilute hydrochloric acid. After a few days the bone can be taken out and washed. The bone is now quite floppy! The acid has dissolved away the non-living part of the bone.

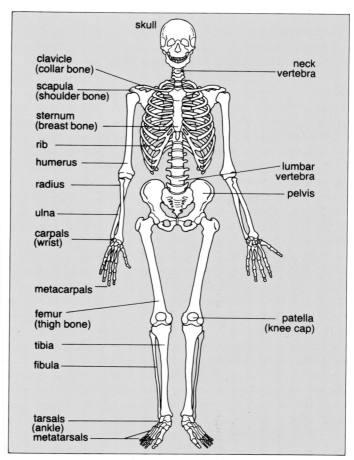

The human skeleton

Activities

1 Try to get a large, fresh bone from a butcher's shop. Don't forget to ask what animal it came from and what part of the skeleton it is.
  a) Draw the bone.
  b) Test the strength of the bone by trying to break it with your hands. (Do not try to smash it with a hammer; splinters of bone may injure you.)
  c) Get someone to cut the bone in half for you using a saw. Draw the inside of the bone. Is the bone hollow? Does it contain marrow?

2 If you have a chicken or turkey for dinner investigate the skeleton.
  a) Find and draw the large, flat bone at the centre of the bird's skeleton. (The large muscles of the wings were fixed to this.)
  b) Find the ribs. What do these protect in the living bird?
  c) Investigate the leg joints.

Moving the machine

The place where two bones meet is called a **joint**. Most of the joints in the body are movable. There are a number of different types. Each one produces a different kind of movement.

Joints are held together by **ligaments**. These are flexible and elastic to allow the joint to move, but strong enough to stop the joint coming apart. When a joint comes apart we say it is **dislocated**. The surfaces of the ends of bones are covered in smooth cartilage and the whole joint is filled with a liquid. This liquid, called **synovial fluid**, acts as a lubricant. The cartilage and the synovial fluid reduce friction and allow smooth movements. Large or uneven strain on joints can damage the cartilage layer. Sportsmen and women need to be careful to avoid sharp, sudden changes of direction if they wish to keep their joints healthy.

Muscles move bones at joints so that the individual bones act as levers. Muscles are attached by **tendons** to bones on either side of a joint. One end is usually attached to a bone which does not move. The other end is attached to a movable bone. When muscles **contract** they get shorter. This gives a pulling force. When the muscles relax they go back to their normal length but they cannot 'push'. Therefore at joints, muscles are arranged in pairs. One muscle of the pair pulls the joint one way and the other pulls it back again. Because the muscles work against each other they are called **antagonistic pairs**.

The human forearm is a good example. The biceps (flexor muscle) bend the arm and the triceps (extensor muscle) straighten the arm.

Questions

1   What is a joint?

2   Why are ligaments   **a)** strong   **b)** flexible?

3   What **two** features of a joint allow smooth movement?

4   Gymnasts sometimes suffer from 'dislocated shoulders'. What do you think this means?

5   Hockey players sometimes suffer from damaged cartilages in their knees. What do you think causes this?

6   What are antagonistic pairs of muscles?

7   Describe what happens to your muscles, joints and bones as you pick up a heavy box from the floor.

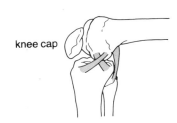

knee cap

Knee joint
This is a hinge joint. The bones can move in one direction only.

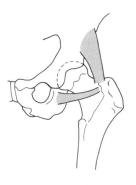

Hip joint
This is a ball-and-socket joint. The bones can move in almost any direction.

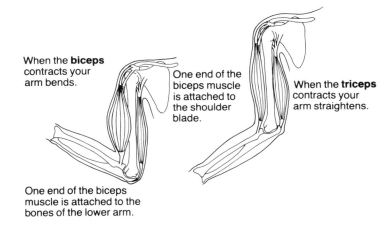

When the **biceps** contracts your arm bends.

One end of the biceps muscle is attached to the shoulder blade.

When the **triceps** contracts your arm straightens.

One end of the biceps muscle is attached to the bones of the lower arm.

Muscles in the upper arm

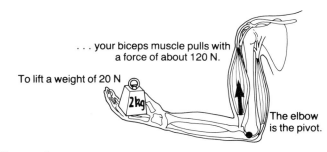

. . . your biceps muscle pulls with a force of about 120 N.

To lift a weight of 20 N

2kg

The elbow is the pivot.

Bones as levers

Keeping the machine going

Like all machines, the human body requires a constant supply of fuel and enough materials for repairs. The fuel for the human machine is **food**. Food supplies energy and all the chemicals needed for growth and to keep the body working properly.

Food types

The things we eat contain different types of food. The body needs these in the right quantities. If one type of food is missing the person can become ill. On the other hand, too much can make a person unfit. For a healthy body you need a **balanced diet**. (See page 17.)

Carbohydrates are chemicals that supply energy. They include sugars, of which there are a number of types, and starch. Carbohydrates release energy when they are used up in cells. **Fats** are also energy foods. They provide more than twice the amount of energy than the same quantity of carbohydrates. A thin layer of fat under the skin acts as insulation and keeps the body warm. Too much fat is not healthy and does not make the body look good.

Proteins are needed to build new body tissues during growth. They are also used to repair damaged tissue. If you cut yourself new skin is made from protein.

Mineral salts are required only in small quantities. The human body needs a wide range of minerals. Calcium is important for the growth of bones and teeth. The haemoglobin (see page 21) in red blood cells contains iron. If you don't get enough iron in your diet, fewer red blood cells will be made. This condition is called **anaemia**. A person suffering from anaemia will look pale and feel tired.

Vitamins like mineral salts are also required only in tiny amounts. However, they play a vital part in the chemical reactions that take place in the body. If a vitamin is missing from your diet you will probably become very ill. Such an illness is called a **deficiency disease**. A deficiency of vitamin C causes scurvy, a disease that makes your gums swell and bleed, and slows down healing of wounds.

Water makes up the large proportion of our bodies. It is essential because all of the chemical reactions of the body take place in solution.

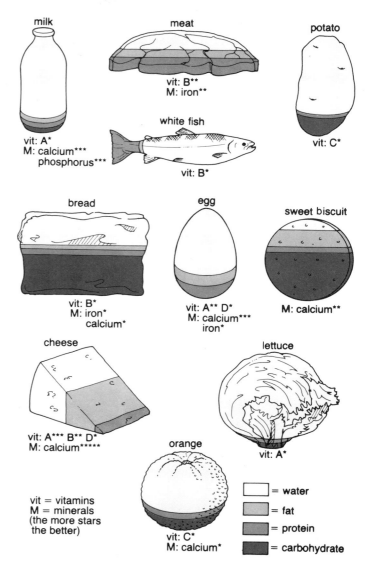

These diagrams show the proportions of carbohydrates, fats, proteins, vitamins and water in some foods.

Questions

1 Which food in the diagram above contains most:
a) carbohydrate **b)** fat **c)** protein **d)** water?

2 A cheese contains 2% carbohydrate, 28% fat and 25% protein. Draw a diagram to show this information.

3 Nearly half the school children in the United Kingdom are overweight. What kind of foods should these children avoid? Explain why.

4 Find out more about anaemia. What causes it? What effect does it have on the body? How can it be cured?

Teeth

Before food can be used by the body it must be broken down into small pieces. The first part of this breakdown is carried out by teeth together with the jaws and the muscles that move them.

There are four types of teeth in the human jaw:

- **Incisors** are shaped like chisels so that they can **cut** cleanly into the food.
- **Canines** are more pointed. We use them for biting but some other animals like dogs have long canine teeth. These are for catching prey and tearing flesh.
- **Premolars** and **molars** have flat, rough crowns which when moved across each other chew and **grind** the food into pulp.

Tooth structure

A tooth has two main regions. The crown, which is the part that you see sticking up from the gum, and the root whose job it is to hold the tooth firmly in its socket in the jaw bone.

Covering the crown is a non-living substance called **enamel**. Enamel is the hardest substance in the human body. It protects the tooth from wearing away. Beneath the enamel is a softer substance called **dentine**. Dentine extends deep into the root where it is covered by cement which holds the tooth in position. In the centre of the tooth is a hollow pulp cavity containing nerves and blood vessels.

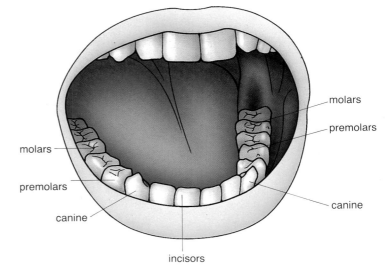

An adult's teeth

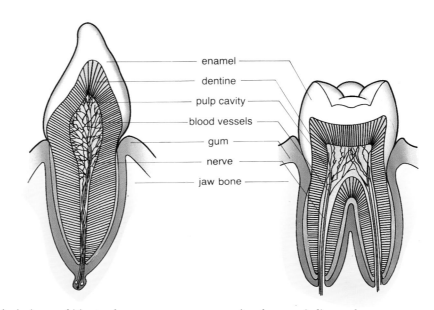

An incisor – a biting tooth   *A molar – a grinding tooth*

Questions

1 The diagram at the top of this page shows an adult's teeth. List the number of   **a)** incisors   **b)** canine teeth   **c)** premolars   **d)** molars. (Note that the top set of teeth matches the bottom set.)

2 Look at **your** teeth in a mirror. Work out the number of each type of tooth in your upper and lower jaw.

3 Name four animals that have got large canine teeth. What do they eat?

4 Name an animal with large incisors. What does it eat?

5 Explain why teeth need to be covered in a hard substance like enamel. (*Hint:* think how many years you will need your teeth!)

Looking after your teeth

Healthy teeth not only enable you to eat all kinds of food but also look good and feel good. Unhealthy teeth increase the chance of infection and can be very painful. Unfortunately few people have a perfect set of adult teeth. **Tooth decay is the most common disease affecting school children in Britain.**

Every time you eat food, some of it remains on your teeth. Bacteria in the mouth feed on this food, particularly if it contains sugar. The bacteria thrive in the warm, moist environment in the mouth and quickly increase in numbers. A yellow material called **plaque** forms on the surface of teeth as colonies of bacteria grow.

The bacteria feed on the sugar and produce an acid. This attacks the enamel, eventually making a hole. Once inside, the bacteria continue to multiply, and feed on the soft dentine. If a dentist finds the damage at this stage he can drill out the decay and put in a filling. If this is not done the bacteria will eat into the pulp cavity and expose the nerves; this causes toothache. If the decay is very bad the tooth may have to be pulled out.

The acid produced by the bacteria can also cause the gums to pull away from the teeth. Bacteria can then get into the gap attacking the tooth and allowing the gums to become infected. Gum disease can cause bad breath. It can also allow the jaw bone to become infected and may make teeth fall out. A good dentist will look for signs of tooth decay and gum disease when checking your teeth.

How to look after your teeth

- Cut down on sweet foods. These help plaque to grow on your teeth.
- Brush your teeth regularly. Thorough brushing with a soft toothbrush will remove plaque **and** massage your gums.
- Use dental floss. This is like a thin waxed cord. It is used to remove food trapped between the teeth.
- Use a toothpaste which contains fluoride. This hardens the enamel and reduces dental decay.
- Visit the dentist regularly. He or she will look for signs of the start of tooth decay or gum disease. These teeth can then be treated so that you do not lose any teeth.

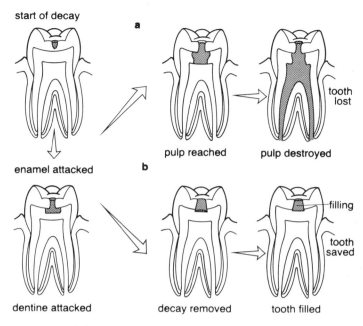

Treating tooth decay

Activities

1 Put a disclosing tablet into your mouth and let it dissolve. Swirl it around your teeth with your tongue for about half a minute.

2 Rinse out your mouth with clean water.

3 Use a dental mirror to see which part of your mouth has been stained.

4 Draw a diagram of your mouth showing as many teeth as you can. Shade those areas that correspond to the stained areas on your teeth.

5 Brush your teeth thoroughly (use the photograph below to help you).

6 Repeat the procedure.

How does cleaning your teeth affect the distribution of plaque?
Which areas of your mouth are difficult to clean?

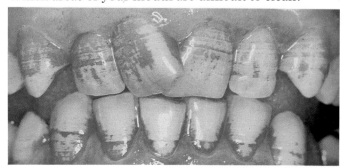

Disclosing tablets show where plaque has built up.

Digesting food

Only very small molecules of food can be absorbed by the body and pass into our blood system. Food, even after a lot of chewing, consists of very large molecules which cannot be dissolved. These must be broken down or **digested** into smaller, soluble molecules.

Digestion takes place in the digestive system and is brought about by the action of chemicals called **enzymes**. (You will read more about enzymes in Chapter 9.)

The enzymes concerned with digestion are mixed with food at various points along the digestive system usually as part of digestive juices.

The digestive system

The gullet has muscles in its walls. These contract and relax to move food into the stomach. This process is called **peristalsis**. *Peristalsis enables you to swallow food even when you are standing on your head!*

In the stomach, food is mixed with **gastric juice** *which contains the enzyme* **pepsin**. *Pepsin begins the digestion of protein. Gastric juice also contains hydrochloric acid which kills the bacteria that are present in food. Continual mixing by the muscular stomach walls produces a creamy liquid which passes into the first part of the small intestine, the* **duodenum**.

Fat, protein and any starch which has not been digested in the mouth, is broken down in the duodenum. **Bile** *is a liquid made in the liver and stored in the gall bladder. It breaks down fats into tiny droplets to form an emulsion.*

The **pancreas** *produces a digestive juice which contains the enzymes that digest starch (amylase), fat (lipase) and protein (trypsin).*

Yet more enzymes are made in the walls of the small intestine. Their action brings about the complete breakdown of food molecules and the end of digestion.

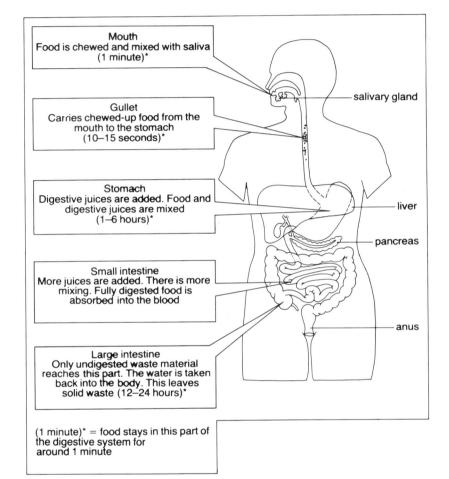

(1 minute)* = food stays in this part of the digestive system for around 1 minute

Questions

1 Why must food be digested?

2 What two jobs are done by saliva?

3 Name *four* parts of our digestive system.

4 Explain what happens to
   a) a starch molecule
   b) a protein molecule
   as it passes through the digestive system.

place	enzyme	food	product
mouth	salivary amylase	starch	maltose (sugar)
stomach	pepsin	protein	polypeptides (small proteins)
duodenum	amylase	starch	maltose
	lipase	fat	fatty acids + glycerol
	trypsin	protein	amino acids
small intestine	maltase	maltose	glucose
	lactase	lactose	glucose + galactose
	sucrase	sucrose	glucose + fructose
	lipase	fats	fatty acids + glycerol
	peptidase	peptides	amino acids

Absorbing digested food

The food which has been broken down during digestion has to be **absorbed** into the blood system before the body can use it.

The digested food particles are so small that they can pass through the wall of the gut (the intestine) by **diffusion**. Diffusion in liquids is a slow process. To make sure that enough food passes into the blood the intestine has a special structure:

- It is long and therefore has a large surface area for absorbing food.

- Its walls are covered with thousands of tiny projections called **villi**. This increases the surface area still further.

- Its lining is very thin, usually only one cell thick, which allows the easy passage of digested food across it and into the blood.

- It is well supplied with blood vessels to carry away food.

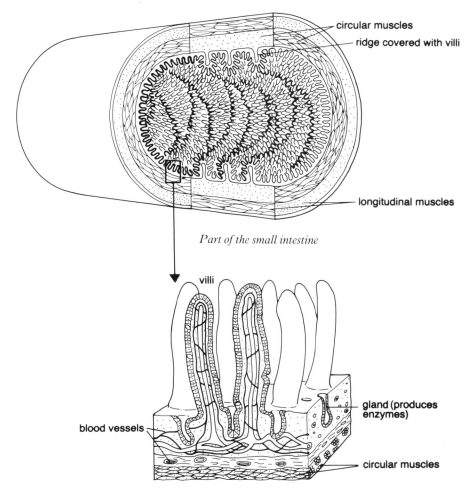

Part of the small intestine

Wall of the small intestine

Food is carried in the blood to the liver. The liver processes food before releasing it to the body for use.

Some material in the food we eat cannot be digested. It consists mainly of cellulose in plant cell walls and is called **roughage**. Roughage is passed into the large intestine where water is absorbed from it before it is expelled through the anus in **faeces**.

Questions

1 Give *four* ways that the intestine is adapted to absorb digested food.

2 Why are these adaptations needed?

3 What happens to food after it is absorbed by the gut?

4 What is roughage?

5 Find out more about the liver. What jobs does it do? Why are these important to the body?

Foods containing roughage

Food tests

You can find out what foods contain by carrying out a few simple chemical tests in the laboratory.

Testing food for starch
Add a few drops of iodine on to the food sample. If any starch is present in the food it will change the colour of the iodine from brown to blue-black.

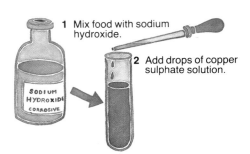

1 Mix food with sodium hydroxide.

2 Add drops of copper sulphate solution.

Testing food for protein
Mix the food sample with about 2 cm³ of sodium hydroxide solution in a test tube. (**Take great care because sodium hydroxide is corrosive**.) *Add a few drops of copper sulphate solution. If protein is present, the colour of the solution will change to purple.*

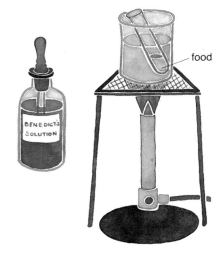

food

Testing food for glucose
Mix the food sample with some water in a test tube and add a few drops of Benedict's solution. Put the test tube in a water bath and heat carefully. (Remember to wear safety glasses.) If glucose is present the colour of the solution will change from blue to green to brick red depending upon the amount of glucose.

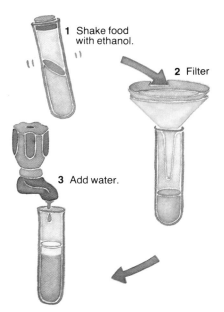

1 Shake food with ethanol.

2 Filter

3 Add water.

Testing food for fat
Thoroughly shake the food sample with some ethanol in a test tube. Filter the mixture into a clean test tube and then add some clean water to the filtrate (the clear liquid which comes through the filter). The presence of fat in the food will be indicated by a white, cloudy emulsion.

Questions

1 Describe how you would find out if
   a) milk contains protein
   b) cheese contains fat
   c) biscuits contain glucose
   d) potato contains starch.

2 What safety precautions should you take when carrying out food tests?

3 Find out what an emulsion is. (*Hint:* the answer is in this book.)

4 Find out what 'Clinistix' and 'Albustix' are used for. (Try your local chemist's shop!)

Healthy eating

Each year a human being eats over 1 tonne (1000 kg) of food and drinks about 500 litres of liquid! Not all of this will be good for maintaining a normal, healthy body. It is important to eat a varied diet – one which provides the right balance between:

- foods for energy
- foods that provide building materials and control chemical reactions
- foods that contain dietary fibre (roughage).

Getting the balance right is not always easy. Many of us just leave it to chance and eat what we like. However, it is clear that in Britain we eat far too much fat, sugar and salt. In addition, we do not eat enough fibre.

Energy foods

Fats and sugars are energy foods. Energy is measured in kilojoules (some people still refer to the old units – calories). The more kilojoules a food contains, the more energy it will provide. If you do not use up the available energy your body stores the excess food as fat and you will become overweight. At the moment about 30% of the adult population in Britain is overweight. Worse still, a lot of school pupils are overweight!

Proteins, vitamins, and mineral salts

Most people get more than enough protein, vitamins and minerals in their normal diet. The body cannot store proteins so eating more will not make you stronger or healthier than you already are. Strength and fitness will only come by carefully balancing healthy eating with exercise. If you are eating enough of the right kinds of foods then you are unlikely to be deficient in vitamins or minerals.

Unfortunately we do eat far too much common salt (sodium chloride). On average we consume about 12 g per day – we could do with half as much. Too much salt leads to high blood pressure so we ought to reduce the amount used in cooking and at the table. More and more food products are now labelled 'Low Salt' or 'No Salt Added' so we have the opportunity to cut down salt intake.

Fibre

Fibre or roughage is made up of the cell walls of plants which pass through the digestive system without being digested or absorbed. It adds bulk to the food giving the muscles in the walls of the digestive system something to push on. Foods containing a lot of fibre help prevent constipation and other disorders of the lower digestive tract such as haemorrhoids ('piles'). We should be eating about 30 g of fibre each day, on average we are only eating half that amount at the moment.

Food additives

Food additives should be listed and their function clearly explained on food packaging, for example, 'preservative – E200 (Sorbic acid)'.

Activities

1 Collect the 'nutritional information' from **four** different food packets or cans.
Which one gives most energy?
Which one contains most fibre?
Which one has the greatest number of vitamins?
Which one has most fat?
Do any of them contain mineral salts? If so, which mineral salts?

Roughage and carbohydrate foods

Protein, vitamins, and minerals

Sugar, starch, and fatty foods

17

Breathing

The air is a mixture of gases, one of which is **oxygen**. Oxygen is needed by all living things so that energy can be released from food during respiration. (You will read more about respiration in Chapter 9.)

During breathing, air is taken into two lungs, oxgyen is removed and carried in the blood to body cells. Carbon dioxide and water, produced in the cells during respiration, leave the body by the reverse process.

Breathing in

The muscles between the ribs contract lifting the rib cage up and out, expanding the chest. At the same time muscles contract to flatten the **diaphragm**. This makes the space inside the rib cage bigger and reduces the air pressure in the lungs. Air moves into the lungs from outside because there the air pressure is higher.

Breathing out

The rib and diaphragm muscles relax. This lowers the chest and raises the diaphragm. The space in the rib cage gets smaller so the air pressure increases. This forces air out of the lungs.

The table shows the amount of different gases in the air we breathe in (**inhaled**) and the air we breathe out (**exhaled**).

gas	inhaled	exhaled
nitrogen	78%	78%
oxygen	21%	17%
carbon dioxide	0.03%	4%
water vapour	varies	saturated

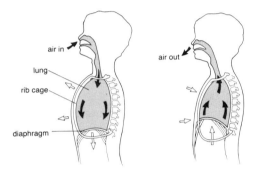

Breathing in   *Breathing out*

When you breathe in the air travels . . .
. . . in through your nose and mouth. . .

. . . down through large tubes called the windpipe and bronchus. . .

. . . into the lungs.
Each lung contains a network of very small tubes. Each tube has an air sac at the end of it.
Air goes along the tubes into the air sacs.
Oxygen passes through the air sac wall. . .

. . . into small blood vessels around the air sac. Blood travels from the lungs to every part of the body. The oxygen goes from the blood. . .

. . . into the cells.
There it is used up.

Carbon dioxide and water go from the cells. . .

. . . into the blood and from the blood. . .

. . . through the air sac walls and into the lungs. . .

. . . up through the bronchus and windpipe. . .

. . . and out through the nose and mouth.

nose
mouth
windpipe
bronchus
heart
lung (outside view)
ribs
lung (inside view)
ribs (the ribs have been cut away to show the lungs)

Lungs in more detail

The lungs are two elastic pouches lying inside the ribs. They are connected to the air outside the body by the windpipe or **trachea**. This opens into the back of the mouth and nose. The trachea divides into two smaller tubes called **bronchi**. One of these goes into each lung before dividing further into smaller tubes called **bronchioles**. After yet more branching the tubes end in tiny, thin walled air sacs called **alveoli**.

Lining all of the air passages are two types of cells. One type is covered with tiny hairs called **cilia**. The other produces a sticky liquid called **mucus**. Small dust particles and bacteria stick to the mucus. The cilia 'beat' to carry the mucus up to the back of the mouth where it is swallowed.

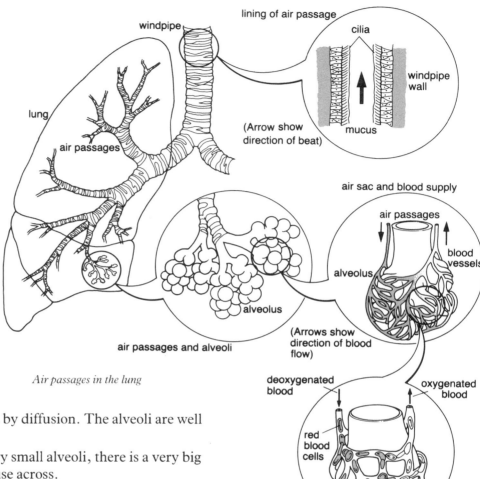

Air passages in the lung

A single alveolus

Oxygen moves into the blood system by diffusion. The alveoli are well adapted to speed up this process.

- Because there are thousands of very small alveoli, there is a very big surface area for the oxygen to diffuse across.
- The walls of the alveoli are very thin – in places just one cell thick.
- The lining of the alveoli is moist so that the oxygen can dissolve.
- The alveoli are surrounded by a dense network of tiny blood vessels called capillaries. These carry the oxygen away.

Activities

Comparing inhaled and exhaled air

The diagram alongside shows a simple piece of apparatus that you can make in school. Tube M is placed in your mouth and then you breathe slowly and gently in and out. The air entering your lungs has to pass through the calcium hydroxide solution (lime water) in test tube A. Air leaving your lungs has to bubble through the calcium hydroxide solution in test tube B.

As you breathe, the solution in test tube B turns a milky white colour. Find out what this shows.

After some time, the solution in test tube A turns a faint white colour. What does this show?

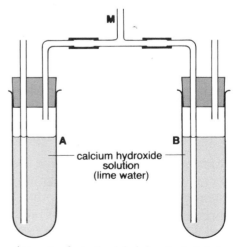

Apparatus for testing inhaled and exhaled air

The air around us is not always as clean as we would like it to be. Not everyone can live high up on mountain sides where the air is fresh and clean. Most of us breathe polluted air which may contain dangerous substances that can damage our health. There are times when it is impossible to avoid breathing in dirty air but some people deliberately breathe in harmful substances! Glue sniffing and smoking are two activities that can seriously damage health.

Glue sniffing

Glue sniffers breathe the vapours given off from certain types of glue and other products in order to get 'high'. The most common products used are the solvents in glue and adhesives, lighter fuel, paint thinners and dry-cleaning agents. These cause mental confusion and hallucinations. They can also cause serious damage to the lungs and brain. Some glue sniffers have died by suffocating on their vomit ('sick') while under the influence of the solvent. Others have suffocated in plastic bags placed over the head whilst sniffing. Most glue sniffers are young people of school age. Attempts have been made to stop shopkeepers from selling glues and solvents to young people.

Smoking

Smoking kills about 50 000 people every year in Britain.

Tobacco smoke contains many chemicals which are harmful to the body.

Nicotine is an addictive drug which is absorbed into the blood stream and seems to affect the heart, blood vessels and nervous system. Nicotine also affects the cilia lining the air passages in the lungs. The natural protection against dust and bacteria is lost and frequent coughing is necessary in order to keep the passages clear. Bacteria infect the air passages making them narrower. This can lead to the difficult and painful breathing associated with **bronchitis** and **emphysema**. Both conditions lead to a reduction of the surface area available for gas exchange and so the sufferers get very 'short of breath'.

Carbon monoxide from cigarette smoke gets into the blood and combines with haemoglobin in the red cells. Oxygen transport around the body is seriously affected. When tobacco smoke cools it forms **tar** which sticks to the lining of the lungs and collects in the alveoli. It is the chemicals in this tar which irritate the lungs and may cause lung cancer.

Smokers affect other people too

Many people who do not smoke find it unpleasant to be in a smoke-filled room. The smoke in the air contains higher concentrations of poisonous chemicals, twice as much nicotine, three times more tar, five times more carbon monoxide. No wonder non-smokers suffer from sore eyes, coughs and headaches. Their clothing smells of cigarette smoke too!

Questions

1 What does the graph show you?

2 What is the death rate from lung cancer for those who smoke 30 cigarettes a day?

3 Explain why smoking can damage your health.

4 Why do you think people take up smoking or glue sniffing?

5 Suggest how young people could be educated so that they do not take up smoking.

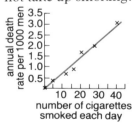

Graph showing the death rate from lung cancer among men who smoke

Getting it around the body (1): blood

What is blood?

Blood consists of a watery, straw coloured liquid called **plasma**. **Red cells**, **white cells** and **platelets** are carried in the plasma. Since there are many more red cells than white cells our blood always looks red. The body of an adult human contains about 5 litres of blood.

Clot formation

Platelets are tiny pieces of cells broken off larger cells in the bone marrow. They play an important part in the clotting of blood. At a cut platelets form a plug preventing continuous bleeding. Gradually a pad of fibres forms beneath the plug and red cells become trapped in it – this is a clot. The surface of a clot hardens off to form a scab.

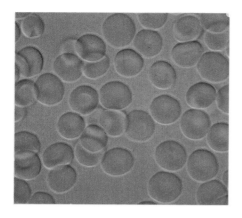

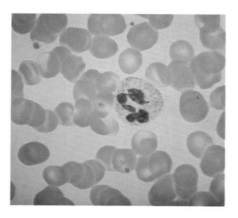

Red blood cells

Red cells transport oxygen from the lungs. They are made in bone marrow and contain a chemical called haemoglobin which makes them red. Oxygen combines with **haemoglobin** *at the lungs to make oxyhaemoglobin. As it travels round the body oxyhaemoglobin slowly releases its oxygen to the cells and changes back to haemoglobin.*

$$Oxygen + Haemoglobin \underset{cells}{\overset{lungs}{\rightleftharpoons}} Oxyhaemoglobin$$

White blood cells

White cells are part of the body's defensive system. Many engulf and destroy any microbes that get into your body through cuts or via your lungs or digestive system. Other white cells produce **antibodies** *which change the poisonous chemicals produced by microbes into harmless substances.*

Blood after spinning in a centrifuge

If blood is spun in a centrifuge the cells and platelets will sink to the bottom of the glass tube leaving the straw-coloured plasma at the top.

Plasma is water in which food, waste material and other substances are dissolved.

What does blood do?

Blood carries chemicals and other substances around the body, that is why the blood and vessels in which it flows is called a **transport system**.

- It carries oxygen from the lungs and food from the digestive system to the cells of the body where it is needed for energy production.
- It carries carbon dioxide from the cells to the lungs where it is removed and breathed out of the body.
- It carries waste materials from body tissues to the kidneys where they are excreted.
- It prevents infection by healing wounds and destroying invading microbes. (You can read more about the prevention of infection on page 35.)

Questions

1 How much blood is there in the body of an adult human being?

2 Why have we got a blood system?

3 What is plasma and what does it do?

4 What differences are there in the shape, structure and function of red and white blood cells?

5 What does haemoglobin do in the body?

The heart is a muscular pump. It pushes blood around the body through tubes called blood vessels.

Blood vessels

Arteries carry blood away from the heart. They have thick muscular walls to withstand the high pressure and usually lie deep inside the body. Arteries divide into small **capillaries** which penetrate into all body tissues so that the blood supply is close to every cell in the body. The transfer of gases, food and excretory products between blood and cells takes place by diffusion through the thin capillary walls. Capillaries join up to form **veins** which return the blood to the heart. Pressure in veins is usually low and valves are present to make sure that blood flows only in one direction.

The heart

The heart is a pump made of muscle. It is probably the best pump ever made since it never stops working from well before you are born until you die. The space inside is divided into four chambers. The top two are the **atria** and the bottom two, the **ventricles**.

Blood from the body enters the right atrium through a large blood vessel called the **vena cava**. Blood from the lungs enters the left atrium through the **pulmonary vein**. The atria contract together pushing blood down into the ventricles. From the left ventricle blood is pumped to the body through the aorta and from the right ventricle to the lungs along the pulmonary artery. Both ventricles contract at the same time.

Every contraction of the atria and ventricles is called a heartbeat. In fact there is really a double beat since the atria contract just before the ventricles. This is important if blood is to flow in the right direction through the heart. There are valves between the upper and lower chambers which prevent any possible backflow of blood.

Your heart beats about 70 times a minute when you are resting but this increases to well over 100 times a minute during physical activity or excitement. You can easily measure your heartbeat by finding your **pulse**. This is a surge of blood produced in the arteries every time the ventricles contract. A pulse can be felt with the finger tips at the wrist where an artery passes over bone close to the surface of the skin.

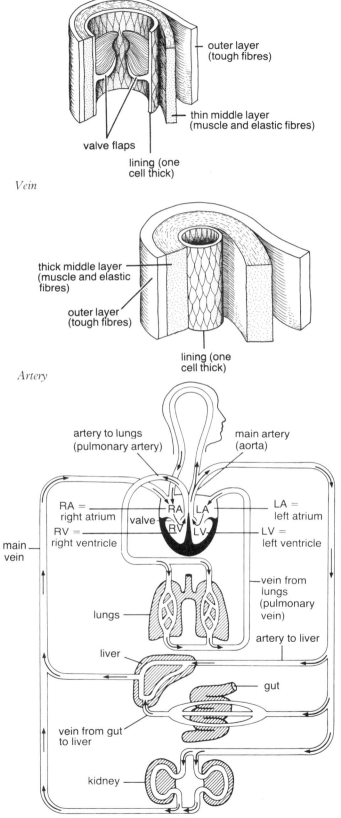

Vein

Artery

The human circulatory system

Coronary heart disease

Coronary heart disease (CHD) kills more people in Britain than any other illness.

The heart like all other muscles in the body requires a good supply of food and oxygen. It gets this through its own blood supply carried in the **coronary arteries**. The coronary arteries are very narrow and as a person gets older, these arteries become 'furred up' with fatty deposits (cholesterol). The arteries become narrower and the blood supply to the heart is reduced. This means that when the heart has to work harder oxygen cannot get to the heart muscle fast enough and a cramp-like pain spreads across the chest. This pain is called **angina** and usually fades after a few minutes rest. Drugs can help relieve the pain, they cause the arteries to widen and allow more blood through.

A heart attack is caused when a coronary artery becomes completely blocked, usually by a blood clot getting jammed in an artery already narrowed by cholesterol. This is called a **coronary thrombosis** ('coronary'). With its blood supply cut off, part of the heart muscle will die causing severe pain or even cardiac arrest where the heart stops beating altogether. Unless the heart can be started again quickly, the person will die.

How to avoid CHD

Medical research has shown that coronary heart disease tends to run in families, it affects older people more than younger ones and men are more likely to suffer from it than women. Obviously you have no control over your family history, age or sex but there are some things you can do to greatly reduce the chances of suffering from heart disease:

- don't smoke at all – it raises the blood pressure
- don't drink too much alcohol – it also raises blood pressure
- eat the right kinds of food – fat (cholesterol) 'furs up' arteries
- exercise regularly to strengthen the heart
- relax more and try to avoid stressful situations.

People who have heart attacks have often ignored these 'rules'.

Treatment for CHD

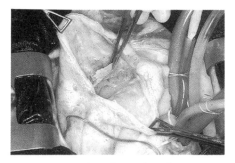

*Over 5000 people each year have **heart bypass surgery**. A piece of vein is removed from elsewhere in the body and used to bypass the blocked coronary artery.*

Heart attack – here I come!

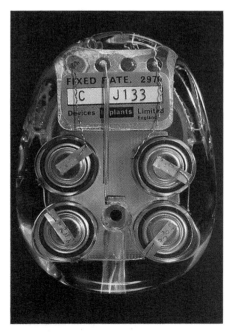

An artificial pacemaker can be implanted if a heart attack destroys the heart's pacemaker (muscle).

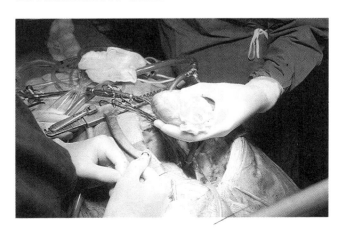

A diseased heart may be replaced completely with a healthy organ taken from another person.

Sight

Sight is perhaps our most useful sense. Try closing your eyes for a few minutes and think of the problems faced by those who are blind.

Eyes are the sense organs of sight. Our eyes are in sockets in the skull. This helps to protect them from damage. The eyes are held in place by muscles. These let us move our eyes from side to side and up and down.

We have two eyes a few centimetres apart. This helps us judge distances and find the position of things easily. The image from each eye is slightly different. The brain puts the two images together to give a 3-D (three-dimensional) effect. This is called **stereoscopic vision**.

How do the eyes work?

Light from an object enters the eye. It is focused by the **cornea** and the **lens** to give a sharp image on the **retina**. The image is small and upside down. The shape of the lens can be changed by a ring of muscle running around its edge. As the ring of muscle contracts or relaxes the lens changes shape. This lets us focus on objects at different distances.

On the retina are millions of light-sensitive cells. Each one is linked to the brain by several nerve fibres. The brain interprets signals from the retina and produces the 'picture' that you see.

There are two kinds of light-sensitive cells – **rods** which are sensitive to light intensity and **cones** which recognise colours.

Rod cells tend to be located around the edge of the retina. Cone cells are found only in the central part of the retina, the **fovea**, and work best in bright light. Have you ever noticed that you cannot see the colour of objects in a dimly lit room?

The amount of light entering the eye is controlled by the **iris**. The muscles of the iris change the size of the **pupil** to allow more or less light to pass through.

Questions

1 Explain why it is an advantage for rabbits to have two eyes, one on each side of the head.

2 What happens to the iris and the pupil when a person in a dark room turns on a bright light?

3 What do the ciliary muscles do to the shape of the lens?

4 Why can't you see coloured objects in a darkened room?

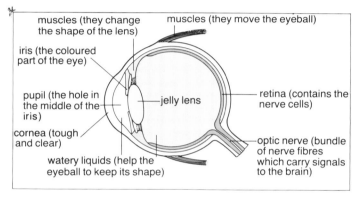

The eye

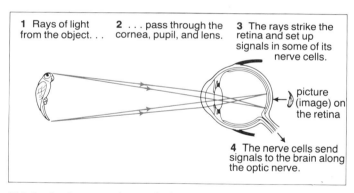

This is what happens when you look at an object.

Activities

1 Hold a pencil at arm's length so that it points upwards. Look at the pencil using one eye. Now, without moving the pencil, look at it using the other eye. You will see that the pencil seems to change position.

2 Hold a pencil in each hand, at arm's length so that they point towards one another. Close one eye and then move the pencils closer together to make their points touch. Repeat the experiment with both eyes open. Which is easier?

Hearing and balance

Hearing

We have one ear on each side of the head. This helps us to tell where a sound is coming from. Ears change vibrations in the air into nerve impulses which travel to the brain where they are interpreted as sound. These vibrations or sound waves are collected by the funnel-like **pinna** and passed down a short canal to the **eardrum**. The eardrum is a thin, tightly stretched membrane which vibrates in time with vibrations in the air.

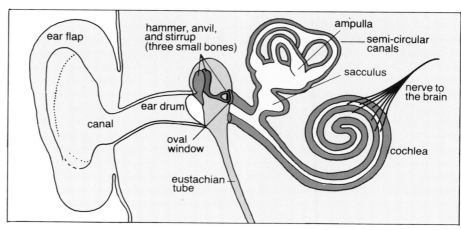

The ear

There is a small air-filled space behind the eardrum connected to the back of the mouth by the **eustachian tube**. This helps to keep the air pressure the same on both sides of the eardrum. Sometimes pressure changes occur outside the ear such as when you take off or land in an aeroplane. This can cause pain. The pressure inside the ear can be equalized by making air enter or leave through the eustachian tube by swallowing or yawning.

Three tiny bones, the hammer, anvil and stirrup, link the eardrum to the oval window. These bones transmit and amplify the vibrations of the eardrum, increasing their force by over 20 times. As a result the oval window is moved in and out sending more vibrations through fluid in a coiled tube called the **cochlea**. Nerve endings in the cochlea detect vibrations in the fluid and signals are sent along a nerve to the brain.

Balance

Your eyes will usually tell you if you are standing upright or not. However, a lot of information about balance comes from structures in the ear called the **sacculus** and **semi-circular canals**.

The sacculus is filled with liquid and lined with sensory hair cells. The hairs of these cells are attached to small, chalk granules. When the head is tilted the chalk pulls on the hairs. The hairs then send signals to the brain where they are translated into information about the movement.

Semi-circular canals are three curved tubes, also filled with fluid, arranged at right angles to each other. Each

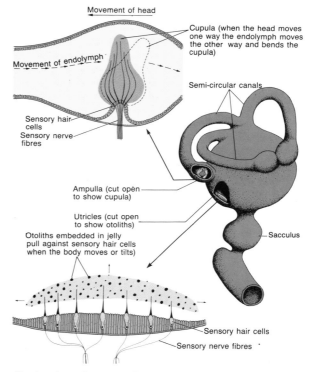

Section through an ampulla

canal has a swelling at the end called the **ampulla**. The ampullae contain structures like those found in the sacculus except that a jelly plate replaces the chalky granules. When the liquid in the canals moves, the jelly plates are pushed to one side and nerve impulses are sent to the brain as before.

The arrangement of semi-circular canals in the head makes it possible for the brain to detect movements in any direction. If you spin around and around then stop suddenly, the liquid in the canals will continue to move against the jelly plates. This confuses the brain and you feel dizzy.

Other senses

Spread throughout the skin are **sense receptors** which are sensitive to touch, pressure, temperature. Some give the sensation of pain.

There are many touch and pressure receptors on the fingertips, lips and tongue. They tell you whether surfaces are hard, soft, rough or smooth. Touch receptors are also attached to the root of every hair on your body.

Small temperature changes in the environment can be detected by temperature receptors of which there are two types, hot and cold. The front of the body has more hot receptors than the back, this explains why your back tends to feel cold when you enter a cold room.

The sensation of pain tells you that something is wrong in the body. Free nerve endings detect pain and these are found all over the skin and inside the body.

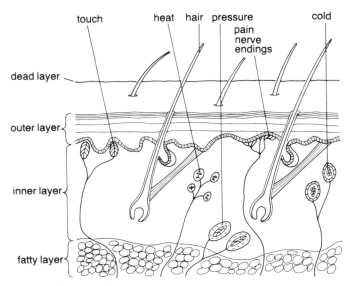

Nerve cells in the skin

Smell receptors in the nose and taste receptors on the tongue both detect chemicals and together are responsible for giving the sensation of flavour in our food.

Olfactory organs at the top of the nose cavity are sensitive to chemicals in the air but only after they have dissolved in the moisture film covering the receptors.

Taste buds are found in between the ridges on the tongue. There are four kinds, each one designed to detect one kind of substance: sweet, salt, sour or bitter. The diagram shows the position of these different taste buds on the tongue.

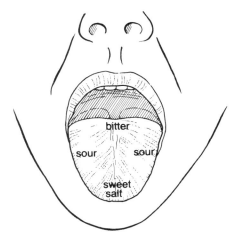

Taste sensors on the tongue

Activities

1 Collect together the following **a)** salt solution **b)** sugar solution **c)** lemon juice (sour) **d)** cold, strong unsweetened coffee (bitter).

2 Using a clean dropper, carefully place a drop of one of the solutions on one of the areas of your tongue shown in the diagram above.

3 Rinse out your mouth with clean water and repeat the experiment using a different solution and different area of tongue.

4 Carry on doing this until you have tested each area of your tongue with all the solutions.

Are your taste buds arranged on your tongue as they are in the diagram? What advantage is there in moving food around your mouth when you are eating? Why do you appear to lose your sense of taste when you have a cold?

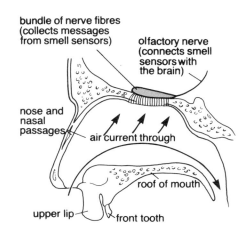

The nose and its smell receptors

Getting information from place to place (1): nerves

All of our body functions must be **coordinated**. This means that they must be made to work together. Coordination is brought about by the nervous system and the hormone system.

The brain and spinal cord together form the **central nervous system** (CNS). Many nerves branch off from the central nervous system to all parts of the body. Making up the nervous system are special cells called **neurones**. There are two types of neurone – **sensory neurones** which carry messages from sense receptors to the CNS, and **motor neurones** which carry messages from the CNS to the muscles and glands. The messages are carried as tiny electrical impulses.

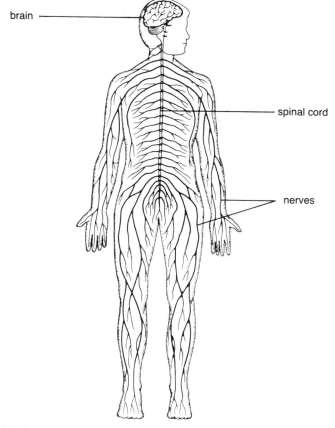

The human nervous system

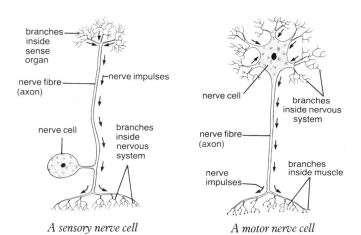

A sensory nerve cell   *A motor nerve cell*

A nerve is a collection of axons from different neurones. It is rather like an electric cable but with an insulation sheath made of myelin rather than plastic.

Nerve impulses pass from one neurone to another by means of a special link called a **synapse**. Branching tips of one neurone lie close to the cell body of another. There is no physical contact between the two. The stimulated neurone releases a chemical that crosses the gap and stimulates the other neurone. Individual neurones can have synapses with many other neurones there are many possible connections that can be made. Perhaps now you realise why we respond to a particular stimulus in a number of different ways.

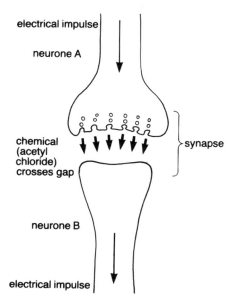

A synapse

Questions

1 What is the central nervous system?
2 Why is it important to be coordinated?
3 What is the difference between a nerve and a neurone?
4 What is a synapse? How does it work?

Drugs and the nervous system

Drugs are powerful chemical substances that act upon the nervous system. They affect the transmission of nerve impulses along individual neurones and from one neurone to another across a synapse. The brain is particularly sensitive to drugs.

Useful drugs

Aspirin and aspirin-based substances are pain-killers. They can be bought over the counter at chemists and other stores usually by people wanting to relieve headaches and other pains. Such drugs should be taken only when needed.

Tranquillizers and barbiturates can only be obtained on prescription from a doctor. Tranquillizers are given to calm people down and make them less anxious. Barbiturates are prescribed to help people sleep – they are sleeping pills. A doctor will closely watch the quantity of such drugs taken by a patient because the body can quite easily become dependent on them.

Some drugs numb parts of the body. These are called **anaesthetics**. Careful use of anaesthetics allows surgeons to carry out operations without causing the patient pain. Dentists often use local anaesthetics to numb your mouth whilst teeth are filled or removed. For more serious operations a general anaesthetic will be used to put the patient completely to sleep.

Harmful drugs

Some drugs produce harmful effects on the body. Unfortunately some people use these drugs to produce pleasant short-term effects. There is a very great danger that users will become dependent or **addicted**. Many drug addicts die either as a direct result of drug taking or from accidents whilst under their influence. **The taking or possession of controlled drugs, such as cannabis, cocaine, and heroin, is illegal.**

The transmission of nerve impulses is speeded up by **stimulant** drugs. Some people take these to stay awake for long periods or to maintain a level of excitement. However, the brain needs to rest and keeping it active for longer than usual will cause damage.

Depressant drugs slow down the activities of the nervous system. They are taken to help people relax and to reduce worry. Unfortunately, since the part of the brain that controls reactions is affected, loss of control and judgement results from their use.

Table: Some of the more common harmful drugs

drug type	name	dangers
stimulants	amphetamines ('uppers' or 'speed')	Addictive – people soon depend on it to maintain a 'high'.
	cocaine ('coke')	Depression and exhaustion when it wears off.
depressants	heroin ('junk' or 'smack')	Makes people lethargic and apathetic. Loss of self control.
	barbiturates ('downers')	Danger of overdose, especially if taken with alcohol, leading to coma and death. Extreme mental and physical deterioration.
hallucinogens	cannabis ('grass', 'marijuana' or 'ganja')	Confusion, lack of coordination mental and nervous breakdown.
	LSD ('acid')	Distortion of space and time. Everything becomes intensified. Serious danger of overdose and accidents.

Hallucinogens alter a person's mood and make things seem larger than life. People talk about 'taking a trip' when taking hallucinogenic drugs. Sometimes the person taking the drug will have a 'bad trip'. They will see horrifying things and become very scared and disturbed. They may even injure themselves in their terror. This type of drug is especially dangerous because of the long term effects it has on the brain. Even after use the effects can recur bringing panic.

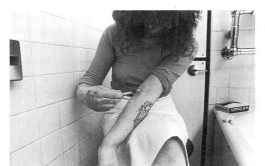

Drug abuse ruins lives

Getting information from place to place (2): hormones

Hormones are chemicals produced by special glands called **endocrine glands**. Like the nervous system, hormones cause various parts of the body to react in different ways but this time the messages travel much more slowly and the effects are more general. Once released into the bloodstream from a gland, a hormone travels all round the body until it reaches **target cells**. The target cells then respond.

There are advantages in having a slow communication system. Hormones can be released into the bloodstream over a long period of time and so control the long-term changes in the body such as growth and development. The liver removes hormones from the blood when they have done their job and are no longer required.

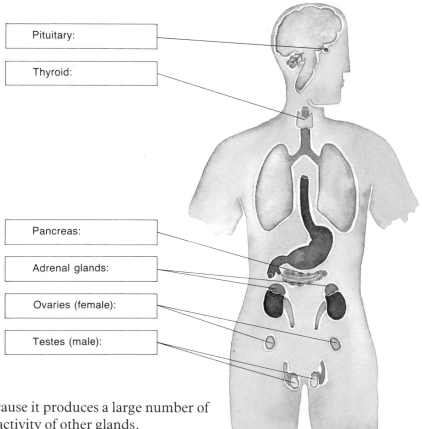

Pituitary:

Thyroid:

Pancreas:

Adrenal glands:

Ovaries (female):

Testes (male):

Pituitary: called the 'master gland' because it produces a large number of hormones, many of which control the activity of other glands.
Thyroid: produces **thyroxine**, a hormone that controls growth.
Pancreas: produces **insulin** which regulates the level of sugar in the blood.
Adrenal glands: produce **adrenalin** in response to stress. This prepares the body for fighting an enemy or for running away by increasing heart beat, breathing rate and blood supply to the muscles.
Ovaries (female only): produce **oestrogen** and **progesterone** which cause body changes at puberty and control the menstrual cycle.
Testes (males only): produce **testosterone** which causes body changes at puberty.

Questions

1 Where are hormones made?

2 How do hormones travel around the body?

3 Explain why nervous 'messages' travel much faster than hormone 'messages'.

4 Diabetes is a disease where the body does not control the blood sugar level properly. Which gland is not working as it should in the body of someone with diabetes?

5 **a)** Which hormone will be released if you are chased by a bull?
 b) Describe two other situations when you might produce this hormone.

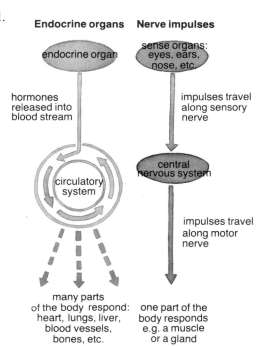

Comparing the hormone and nervous systems

Reproduction (1): males and females

The time when boys and girls become sexually mature is called **puberty**. Puberty usually starts at the age of 11–13 in girls and about 12–14 in boys. People vary a great deal and so there is nothing wrong with you if puberty comes earlier or later in your life.

What happens at puberty?

As well as growing fast (perhaps 15 cm or more in a year), at puberty we develop secondary sexual characteristics. In women these include breasts and a more rounded figure. Men grow more facial hair and develop a more muscular body and a deeper voice. Hair grows under the arms of both sexes and around a woman's vagina and a man's penis.

Puberty is the time when reproductive organs start to produce **gametes**. These are the sex cells which are involved in reproduction.

At puberty, girls start to produce mature eggs in their ovaries. They will notice this because each month they will lose a small amount of blood through the vagina. This is known as **menstruation** (see page 33).

Boys start to produce sperms in their testes. Occasionally these sperms will be ejected with seminal fluid during the night. These 'wet dreams' are completely natural and quite harmless.

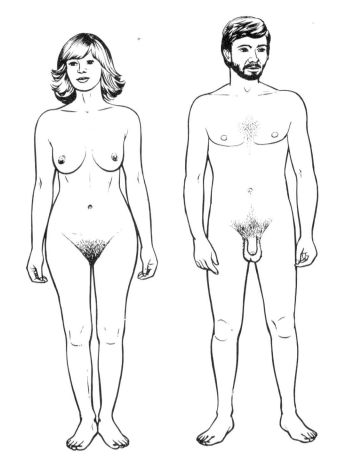

Male and female bodies

Female reproductive organs

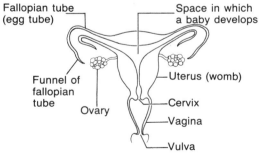

Fallopian tube (egg tube)
Space in which a baby develops
Funnel of fallopian tube
Ovary
Uterus (womb)
Cervix
Vagina
Vulva

The ovaries contain thousands of potential egg cells when a girl is born. However, only a very small number will ever mature later in life.

Male reproductive organs

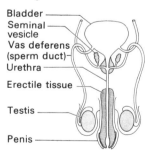

Bladder
Seminal vesicle
Vas deferens (sperm duct)
Urethra
Erectile tissue
Testis
Penis

The lining of the tubes which make up the testes consists of cells which are constantly making sperms. Billions of sperms will be made during a lifetime.

Questions

1 Where are male sex cells made?

2 What are female sex cells called?

3 Explain why a boy does not need to shave until he reaches his mid-teens.

4 Describe the changes that take place in the body of **a)** a boy **b)** a girl during puberty.

5 Girls can sing high notes all through their life but boys are unable to do so after the age of about 14. Why is this?

Reproduction (2): fertilization

An egg is fertilized when the nucleus of a single sperm fuses with the egg nucleus.

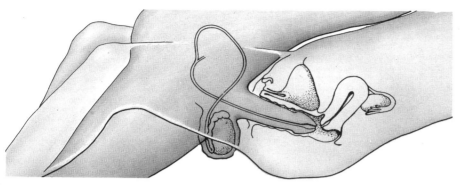

Sexual intercourse

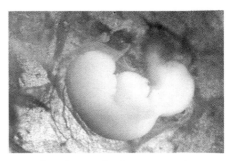

. . . at 2 weeks eyes develop, arms and legs are just bumps.

For sperm to meet an egg, **mating** or **sexual intercourse** must take place. During sexual excitement the man's penis becomes enlarged and stiff (erect) due to an increase in blood supply. This erection makes it possible for the penis to be pushed into the woman's vagina. The sensitive tip of the penis is stimulated by being moved up and down the vagina. After a while, muscular contractions of the sperm tubes **ejaculate** sperms into the vagina. The good sensation felt by the man during ejaculation is called an orgasm. A woman may also experience an orgasm during sexual intercourse. About 400 million sperms are released by the man during each ejaculation. The sperms are mixed with seminal fluid. Seminal fluid contains chemicals that nourish the sperm and encourage them to start swimming. Sperm travel through the womb and up the oviducts where an egg may be waiting. If the egg is fertilized it divides repeatedly to form a ball of cells which passes down the oviduct and attaches itself to the prepared wall of the womb.

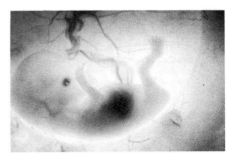

. . . at 6 weeks the heart begins to beat and hands and feet begin to grow.

3 Fertilization takes place when a sperm joins with the egg as it moves down the tube.

4 The fertilized egg is moved down the tube into the womb. The cells keep dividing.

5 To prepare for the egg, the womb has been growing a new lining full of blood vessels.

2 The egg moves down the tube.

1 A fully developed egg leaves the ovary and enters the egg tube. At the same time sperm cells may travel up from the vagina.

ovary

muscles round the womb

6 The embryo becomes fixed to the lining of the womb.

Fertilization of an egg

The cells in the ball divide over and over again to produce tissues and organs. During the next 9 months the developing baby, or **embryo** grows inside its mother. This is called the **gestation period** or **pregnancy**.

Pregnancy usually lasts for about 38 weeks. After this period of time the baby is ready to be born.

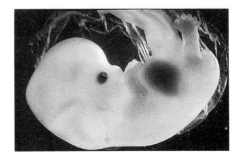

. . . at 12 weeks the baby begins to move its arms and legs. Fingers and toes are formed.

Reproduction (3): birth

During pregnancy the womb will have become thickened with extra cells and supplied with an increased blood supply. Connecting the embryo to the womb wall is the **umbilical cord** and **placenta**. These are formed from the tissues of the embryo as it develops. A bag, the **amnion**, filled with a liquid called amniotic fluid surrounds and protects the embryo throughout pregnancy.

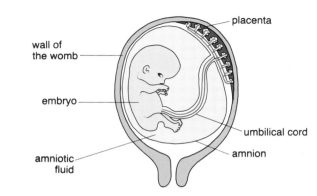

An embryo in the womb

The placenta extends into the womb wall and substances dissolved in the mother's blood pass across it into the baby's blood system. It controls the entry of materials into the baby's blood, keeping out many harmful substances and allowing essential ones like food and oxygen in. Unfortunately alcohol, nicotine and some drugs can pass through this barrier. This is why a pregnant woman should avoid drinking alcohol and smoking cigarettes. She should also take drugs only when they have been prescribed by a doctor.

The blood system of mother and baby never actually mix. The mother's blood pressure would damage the developing circulatory system of the baby and if the blood groups were different it could cause clotting.

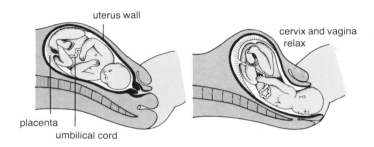

The baby ready for birth   *The birth*

A few weeks before birth the baby usually turns upside-down so that its head is just above the neck of the womb. 'Labour pains' start when the womb begins to contract rhythmically – slowly at first, then quicker and stronger. Muscular contractions of the womb break the amnion and fluid is released. The neck of the womb relaxes and baby's head passes through. Eventually contractions of the womb together with those of the abdomen muscles, push the baby through the vagina and out into the world.

During the first months of its life outside the womb the baby will get all the food it needs from its mother. It suckles milk produced in the **mammary glands** inside its mother's breasts. As it gets older the growing child will begin to eat solid food.

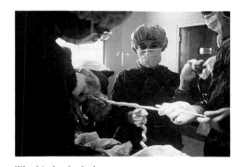

The birth of a baby

Questions

1 What is meant by the following words
   a) erection   **b)** ejaculation   **c)** fertilization?

2 What happens to the penis before sexual intercourse?

3 Describe what happens to a sperm from ejaculation to fertilization.

4 Describe what happens to an egg after fertilization.

5 What is the placenta for?

6 What are 'labour pains'?

7 Try to find out how long pregnancy lasts in   **a)** a mouse
   b) a dog   **c)** a human   **d)** an elephant. What pattern can you see?

Reproduction (4): menstruation

Many hundreds of eggs are produced by a sexually mature woman. These are released approximately one per month. It is very unlikely however that more than a very small number of these eggs will ever be fertilized either because sexual intercourse hasn't taken place or it took place at a time when an egg wasn't present in an oviduct.

Eggs only live for a short time, about 48 hours. When an egg is not fertilized the thickened lining of the womb disintegrates. Unwanted cells together with some blood are pushed from the womb. The waste material is moved out of the body through the vagina by contractions of the womb wall. This monthly loss of blood is called menstruation or a **period**. A period lasts for about five days and can cause some discomfort. This is because the contractions of the womb wall can cause cramp-like pains. These pains can usually be treated with pain-killing tablets like paracetamol.

To absorb the blood lost during menstruation women wear sanitary towels or tampons which must be changed regularly.

Many women get depressed and irritable during the days before menstruation. This is often referred to as **pre-menstrual tension** or PMT.

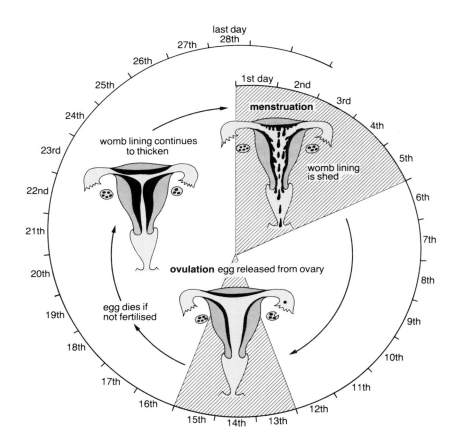

The menstrual cycle

Questions

1 If a woman begins to produce eggs at the age of 13 and stops at 48, roughly how many eggs will she produce during her life?

2 During one menstrual cycle a woman's period started on March 16th. On what date would an egg be released from an ovary? (The diagram will help you.)

3 What changes take place in the wall of the womb as it prepares to receive a fertilized egg?

4 What changes take place in the wall of the womb if it does not receive a fertilized egg?

5 What is the purpose of menstruation?

6 Why do you think menstruation stops when a woman is pregnant?

Contraceptives and STD's: things you should know about

Sexual intercourse is pleasurable and is a natural extension of the love between a man and a woman. However, if the couple do not want to have a child they must somehow stop a sperm from fertilizing an egg. Fertilization can be prevented by the use of contraceptives. There is a number of types:

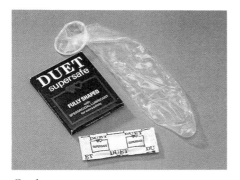
Condom

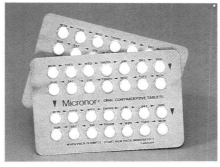
Contraceptive pills

Coil

What are STDs?

STDs are **sexually transmitted diseases**. These are passed on during sexual activity. They affect the reproductive organs mainly but can spread to other body parts. Some of the more common STDs include Gonorrhoea, Syphilis, Genital Herpes, NSU (Non Specific Urethritis), TV (Trichomoniasis Vaginalis), Thrush (Candidiasis) and lice or 'crabs'.

The microbes causing an STD cannot live for long outside the body therefore it is very unlikely that they can be caught by anything other than sexual intercourse.

AIDS

AIDS (Acquired Immune Deficiency Syndrome) is a disease which damages the body's natural defensive system. People with AIDS are unable to fight off infections that the body would normally resist. Such infections are usually fatal.

AIDS is not easy to catch. The microbe causing it, a virus, cannot live outside the body so AIDS cannot be spread by coughing, sneezing or drinking from the same cup. The virus is passed from one person to another when blood, semen or vaginal fluids are mixed. This may happen during sexual intercourse or when injecting drugs with used needles or syringes.

There is as yet no known cure for AIDS. So it is important to follow some simple rules if you are to reduce the risk of infection.

- Have few sexual partners. Unless you are sure of your partner always use a condom.
- People who use drugs should never share syringes, needles etc.
- If you have your ears pierced or body tattooed always make sure that equipment is sterilized first.

The Condom
The condom is a thin rubber sheath that is fitted over the erect penis before intercourse. Ejaculated sperm is collected in a small bulb at the tip. Though not perfect as contraceptives, condoms have no physical side effects and are the only method of birth control that help prevent STDs.

The Pill
The pill contains a delicate balance of hormones that prevent ovulation. The pill is prescribed by a doctor and must be taken exactly as instructed. Though regarded as a reliable contraceptive, there are possible side effects such as weight gain.

The Coil
The intra uterine device (IUD) is a small coil or loop of plastic inserted by a doctor into the womb. Its presence in the womb appears to prevent implantation of the fertilised egg. IUDs are not 100% effective and sometimes there are side effects such as very heavy periods.

Microbes and disease

Many infectious diseases are caused by germs which are types of **microbes.** Microbes include viruses, bacteria, single-celled organisms called protozoa and some fungi.

Some diseases caused by microbes:

microbe	disease
virus	measles mumps genital herpes
bacteria	whooping cough pneumonia gonorrhoea
protozoa	dysentery sleeping sickness trichomoniasis
fungi	athlete's foot ringworm thrush

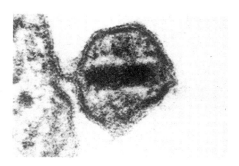

Viruses are extremely small and can only live inside the cells of other living things. They take over control of the cell contents instructing them to make more of their own kind.

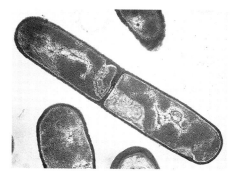

Bacteria are larger than viruses though still very small. There is a great variety of bacteria; some are very useful such as those that decompose dead organisms; others can be very harmful.

Once harmful microbes get into your body, through a cut for example, they reproduce rapidly. They feed on your body cells and produce poisonous waste called **toxins**. As the numbers increase, so more cells are damaged and more toxins are produced – you soon begin to feel ill.

Earlier in this chapter you read about white blood cells and their role in protecting the body against infection by microbes. Certain types of white cells, called **phagocytes**, engulf and digest microbes. Others produce **antibodies** which 'glue' microbes together making it easier for phagocytes to do their job. Yet another group of white cells produce **antitoxins** which neutralize the toxins.

When you recover from a particular disease antibodies remain in your blood system for only a short while. However, the ability to produce them in increased. This is very useful because if your body is invaded by the same kind of microbe again you are able to fight them off rapidly and so not suffer the disease. You are said to be **immune** to that disease.

Immunity to diseases can be given by **vaccination**. A vaccine is a liquid containing dead or weak microbes. When it is injected into your body you get a very mild form of disease, so mild that you probably never notice it. However, the injection is sufficient to stimulate the production of antibodies and bring about immunity to the disease.

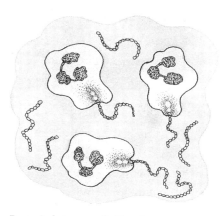

Bacteria being engulfed by phagocytes

Questions

1 Make a list of the diseases you have had. Which ones have you had more than once? Why do you think you can catch some diseases more than once?

2 Find out what diseases you have been vaccinated against. Explain what has happened inside your body as a result of these vaccinations.

3 A woman going to work in Africa had a vaccination against the serious disease cholera. She felt weak and ill for about 12 hours after the injection but then began to feel much better. Why do you think the injection made her feel ill? Why was it important for her to have the injection of vaccine?

Cancer

Cancer is a disease in which cells multiply in an uncontrolled way. The cells multiply so quickly that a lump or **tumour** forms. The cancerous cells then grow out and destroy healthy parts of the body. Some cells may even break away and be carried in the blood to other parts of the body where they cause new tumours. The uncontrolled cell growth uses up the body's resources so cancer patients may feel tired all the time and lose a lot of weight. If the cancer is not successfully treated, the victim will die.

Not all lumps that appear in the body are due to cancer. Fatty lumps, warts and cysts are usually perfectly harmless. They can be removed by surgery quite easily. Tumours which do not spread or cause damage are called **benign tumours**. Tumours which spread and destroy healthy tissue are called **malignant tumours**. Many malignant tumours can be successfully removed by surgery **if they are found at an early stage**.

No one knows exactly what causes cancer but some chemicals, asbestos fibres and radiation appear to make cancer more likely. Chemicals that cause cancer are called **carcinogens**. We know many chemicals that seem to be carcinogenic and others are being discovered all the time.

The medical world now agrees that the tar in cigarette smoke contains carcinogens. This means that cigarette smokers are much more likely to get lung cancer than non-smokers. In fact about 29 000 people die from lung cancer in Britain each year. Approximately 26 000 of these are smokers!

Many cancers still prove fatal but modern research is finding new ways of treatment all the time. New drugs and surgical techniques cure many people. Others are cured by radiation treatment (see page 103).

Allergies

An **allergy** is a condition where a person's body reacts badly to a substance which is usually harmless to others.

There are many types of allergy, in fact people can be allergic to almost anything including household dust, wool, cosmetics and certain foods like eggs, shellfish, even beer! Perhaps the most common allergy is **hay fever**. This is a reaction to pollen from plants. The amount of pollen in the air (the pollen count) is greatest in the summer. Allergies cause the production of a chemical called **histamine**. This causes swelling, redness and itching of surrounding tissues. If this occurs in the nose and around the eyes all the unpleasant symptoms of hay fever appear – a runny, itchy nose and sore, watery eyes.

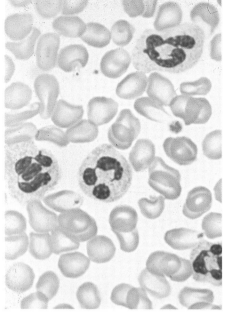

Healthy cell

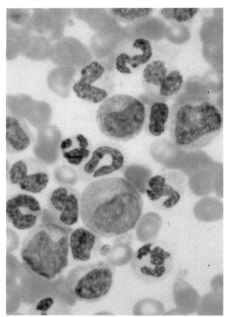

Cancerous cell

Questions

1 The table (right) was printed on a packet of breakfast cereal.
   a) i) How many vitamins are there in the cereal?
        ii) Why do we need vitamins?
   b) i) Why do we need dietary fibre in our food?
        ii) Explain why these cereals are advertised as 'high fibre'.
   c) A man eats 50 g of this cereal.
        i) How much energy will he get?
        ii) How much protein will he get?

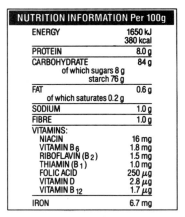

NUTRITION INFORMATION Per 100g	
ENERGY	1650 kJ 380 kcal
PROTEIN	8.0 g
CARBOHYDRATE	84 g
of which sugars 8 g starch 76 g	
FAT	0.6 g
of which saturates 0.2 g	
SODIUM	1.0 g
FIBRE	1.0 g
VITAMINS:	
NIACIN	16 mg
VITAMIN B$_6$	1.8 mg
RIBOFLAVIN (B$_2$)	1.5 mg
THIAMIN (B$_1$)	1.0 mg
FOLIC ACID	250 μg
VITAMIN D	2.8 μg
VITAMIN B$_{12}$	1.7 μg
IRON	6.7 mg

Nutrition label from a packet of cornflakes

2 The pulse rate and breathing rate of a young person during various activities is shown in the following table:

activity	pulse rate/min	breathing rate/min
sleeping	60	8
eating	70	17
walking to school	90	23

   a) Explain why the breathing rate is different for each activity.
   b) Explain why the pulse rate is different for each activity.
   c) What would you expect the pulse rate and breathing rate to be immediately after vigorous exercise?
   d) Describe how you would measure pulse rate.

3 The diagram shows antagonistic muscles in the human forearm.
   a) What are antagonistic muscles?
   b) Why do we need two muscles to bend and straighten our arm?
   c) Explain what happens when you move your right hand upwards to touch your right shoulder.

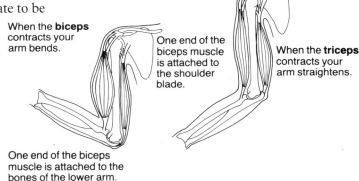

When the **biceps** contracts your arm bends.

One end of the biceps muscle is attached to the shoulder blade.

When the **triceps** contracts your arm straightens.

One end of the biceps muscle is attached to the bones of the lower arm.

4 'The human blood circulatory system performs the function of transport and defence.'
   a) Name two things that are transported in the blood.
   b) What makes blood circulate in the body?
   c) Explain how blood defends the body from illness.
   d) What is AIDS? Why do people die from AIDS?

5 **a)** Describe what happens to a boy or girl at puberty.
   b) Describe what happens from the start of sexual intercourse to the successful fertilization of an egg.
   c) Name one sexually transmitted disease. Describe its cause, its symptoms and how it can be treated.

6 The diagram below shows one of the tiny air sacs (alveoli) in a lung.

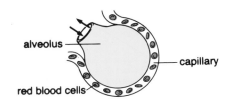

alveolus
capillary
red blood cells

   a) What is a capillary?
   b) What gas diffuses from the alveoli into the capillaries?
   c) Name one gas which diffuses from the blood into the alveoli.
   d) Why is the surface of each alveolus i) moist ii) thin?
   e) How does smoking damage the lungs?

2 Motion

Why study movement?
What are speed and acceleration?
How do we make things move?
How do we make things stop?

Wherever we look in the universe we find things that are moving. Animals, including humans, are clearly able to move and we can build 'machines' such as cars, boats and aeroplanes for transport. However, we also find motion in less obvious places.

On a very small scale, the tiny particles which make up atoms are always on the move.

At the other end of the size scale, huge galaxies are made up of millions of stars moving around. Even our own planet orbits the Sun in a regular pattern.

By studying moving things, scientists can find laws or patterns which 'fit' the movement or motion. This means that they can predict how other things will move.

Car designers use these laws to improve the performance of new models. Aerospace engineers use them to calculate how to put rockets into space and satellites into orbit. Astronomers can find out more about the universe by observing and measuring how stars move. Athletes can even improve their performances by studying the science of motion.

There is no end to the use which we can make of the laws of motion.

Speed

What do we mean by speed?

Speed tells us how fast a thing is moving. For example, the speed of a car tells us how fast it is moving by telling us how far it will move in a set time. Look at the car speedometers drawn below. Car 1 is travelling at a speed of 100 km per hour (100 km/h). If it stays at this speed it will travel a distance of 100 km in one hour.

  Car speedometers

Car 2 is travelling more slowly at 80 km/h. At this speed the car travels just 80 km in one hour.

From the speed and the time it is easy to work out how far the car will travel. The table opposite gives some examples for cars 1 and 2.

What do you notice about the numbers?

We can say that;

$$\text{distance} = \text{speed} \times \text{time}$$

We can use symbols;

$$s = vt$$
$$(s = \text{distance}, \ v = \text{speed}, \ t = \text{time.})$$

The speed is also useful because we can use it to work out how long a journey will take.

The table opposite gives some examples for cars 1 and 2. What do you notice about the numbers?

car	speed	time	distance travelled
1	100 km/h	½ h	50 km
1	100 km/h	1 h	100 km
1	100 km/h	2 h	200 km
2	80 km/h	½ h	40 km
2	80 km/h	1 h	80 km
2	80 km/h	2 h	160 km

car	length of journey (distance)	speed (average)	time taken
1	200 km	100 km/h	2 h
1	350 km	100 km/h	3½ h
2	40 km	80 km/h	½ h
2	200 km	80 km/h	2½ h

We can say that;

$$\textbf{time taken} = \frac{\textbf{distance travelled}}{\textbf{speed}} \text{ or, in symbols, } t = \frac{s}{v}$$

Miles per hour (mph) and kilometres per hour (km/h) are very useful units for measuring the speed of things like cars and trains. For other things it is often useful to measure the speed in metres per second (m/s).

We can still use the same 'laws';

$$\text{distance} = \text{speed} \times \text{time} \qquad \text{time} = \frac{\text{distance}}{\text{speed}}$$

Questions

1 A sports car travels at its top speed of 200 km/h. How far will it travel in **a)** 5 hours **b)** 30 minutes?

2 A train travels at 180 km/h. How far will it travel in 4 hours?

3 A snail travels at 50 cm/h. How far will it travel in **a)** 3 hours **b)** 1.5 hours?

4 A sports car travels at its top speed of 200 km/h. How long will it take to travel
a) 400 km **b)** 50 km?

5 Concorde travels at 2300 km/h. How long will it take to travel from London to New York (6900 km)?

6 A snail travels at 50 cm/h. How long will it take to cross a flower bed which is 125 cm wide?

7 A bullet from a gun travels at 600 m/s.

How far does the bullet travel in 2 s?

How long would it take to travel
a) 300 m **b)** 150 m?

8 A sprinter can run at an average speed of 10 m/s. How long does he take to run **a)** 100 m **b)** 60 m? Why do you think that the athlete cannot run 800 m in 80 s?

9 In air, sound travels at around 300 m/s. How long does it take to travel 1500 m?

10 A goal keeper shouts at a footballer 100 m away. How long does it take the goal keeper's message to arrive?

Measuring speed

We have seen that if we divide the distance travelled by something which is moving by the time taken to cover that distance, we get the speed. When measuring speed we can use this formula;

$$\frac{\text{distance}}{\text{time}} = \text{speed}$$

Activities

Measuring the speed of cars

For things like cars we can measure how much time they take to travel a fairly large distance.

Choose a road where cars travel at almost steady speeds. Why would near a corner or close to a set of traffic lights not be good places?

Measure the distance between two lamp posts or other fixed objects. These markers should be about 50 m apart.

A person with a stop watch should stand at the first marker. Another person stands at the second marker.

As a car passes the person doing the timing starts the stop watch. As the car passes the second person he or she must signal for the watch to be stopped. It will take practice before you can time accurately.

Some sample results are given below. A calculator helps!

SAFETY

1   Keep well clear of the road. You should not stand on the kerb.

2   Take care not to distract drivers with your signals.

Car	Distance between markers (in m)	Time Taken (in s)	Speed (in m/s)
1	65	5·0	65/5·0 = 13·0
2	65	4·6	65/4·6 = 14·1
3	65	3·2	65/3·2 = 20·3

Sample results

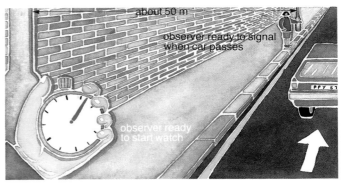

about 50 m
observer ready to signal when car passes
observer ready to start watch

Investigating car speeds

Using your results

After measuring the speed of about fifty cars you should be able to answer these questions;

a) What was the fastest that anyone drove during your investigation?

b) What was the 'average' speed of the cars in your survey?

c) What fraction of the cars were breaking the speed limit? The conversion chart below may help.

speeds		
mph	**km/h**	**m/s**
30	48	13
40	64	18
50	80	22
60	97	27
70	113	31
80	129	36
90	145	40
100	161	45
All to the nearest whole number.		

Questions

1   In an Olympic swimming competition, an electric clock starts when the starter's gun is fired. The clock is stopped when the swimmer touches the end of the pool.

Work out the average speed for each of these swimmers.

  a) Backstroke . . . 100 m in 60 s
  b) Breaststroke . . . 200 m in 140 s
  c) Butterfly . . . 200 m in 120 s
  d) Freestyle . . . 400 m in 230 s

Measuring speed with a ticker-timer

The ticker-timer is often used in school laboratories to measure speeds. It uses mains electricity to make a flexible metal strip move up and down fifty times in each second! Each time it moves down it presses on a piece of inked paper ('carbon' paper) and leaves a dot on a piece of paper tape.

If the paper tape is not moving, in one second the ticker-timer makes fifty dots **all** in the same place. However, if the tape is fixed to something which is moving, the tape is pulled through and so the dots are spread out. We can work out the speed by measuring how far the tape has been pulled through. We can find the time from counting the dots on the tape.

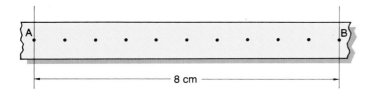

This tape was pulled through the timer at a steady speed. We can see that the dots are all the same distance apart. There are ten 'gaps' between the lines drawn so the tape took one fifth of a second to make (0.2 s). The distance between the lines is 8 cm.

$$\text{Speed } v = \frac{s}{t} \qquad \frac{8\,\text{cm}}{0.2\,\text{s}} = \textbf{40 cm/s}$$

Looking at changing speeds

The useful thing about a ticker-timer is that the dots on the tape show us whether the speed changed during our measurements. Think back to our car speed activity. We did not measure whether the car speeded up or slowed down between the lamp posts. We worked out the speed as if it stayed steady (constant).

On the tape from a ticker-timer, if the dots get further apart we know that the object was speeding up. If they get closer together then the object was slowing down.

We can cut the tape into strips with equal numbers of dots. By sticking them side by side, we can make bar charts. We can think of these bar charts as graphs showing how the speed changes with time.

The tapes show three different types of motion.

Ticker-timer and tape

Equal distances → steady speed ('constant speed')

Dots get further apart → getting faster ('accelerating')

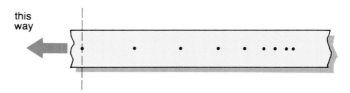

Dots get closer together → slowing down
('decelerating')

Questions

1 a) Work out the speeds shown by these tapes.

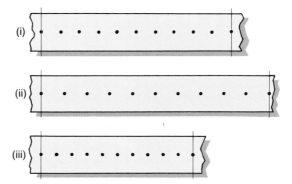

b) What do you notice about the distance between dots and the speed?

Speed-time graphs

When we make 'real' journeys our speed changes many times. Think of a train journey. When you get on the train it is not moving. As it leaves the station it gains speed or accelerates. On a long straight track the train can keep to a constant speed. The driver will need to change the speed as the train approaches bends, level crossings and stations. We can show all these variations on a speed–time graph.

The graph below shows the speed of a car during part of a journey through a town. We can tell several things from the graph.

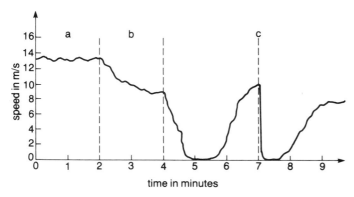

Speed-time graph for a car journey

a) For the first two minutes the driver was driving at an average speed of 13 m/s. During the next two minutes the speed was lower. Perhaps there was more traffic.
b) At a time of four minutes, the driver slowed down and then stopped. The car did not move for almost a minute. Perhaps the driver had to stop at a pedestrian crossing.
c) At a time of seven minutes, the car stopped very suddenly. Perhaps a child ran out into the road and the driver had to do an emergency stop.
d) After about 30 seconds, the car moved off again. Obviously there was no accident after the emergency stop.

One of the things we can work out from a speed–time graph is the distance which the moving object has travelled.

The area under a speed–time graph is equal to the distance travelled.

This rule works for any shape of speed-time graph. The example is for an aircraft stopping after landing.

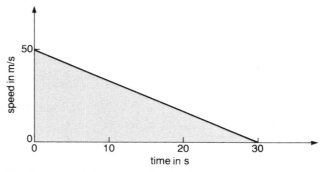

Speed-time graph for an aircraft on landing

The area of any triangle is given by half its base length multiplied by its height. For this triangle the area is:

$$(\tfrac{1}{2} \times 30\,\text{s}) \times 50\,\text{m/s} = 15\,\text{s} \times 50\,\text{m/s} = \textbf{750 m}.$$

The runway must be at least 750 m long for this aircraft to land safely.

Questions

1 The speed–time graph below shows (roughly) the speed of a sprinter in a 100 m race.

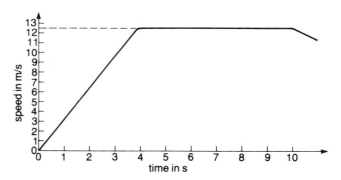

Speed-time graph for a sprinter

a) How long did it take for the sprinter to reach top speed?
b) What was his top speed?
c) What was the sprinter's speed after 2 s?
d) What was the sprinter's speed after 1 s?
e) How far had the sprinter run after 4 s? (*Hint:* the area under the graph is a triangle).
f) How far had he run after 10 s?

Changing speed and direction

The modern study of motion owes much to Sir Isaac Newton. He was an English mathematician and scientist who, amongst other things, worked to find laws which would explain the movements of planets and things on Earth. He is remembered now for the 'laws of motion' which he set out.

Newton's first law says: 'objects will continue to travel in the same direction with the same speed unless an unbalanced force acts on them.'

What do we mean by unbalanced forces?

Forces can be thought of as 'pushes' or 'pulls'. The size of a force is measured in units called **newtons**.

When an object is pulled by two forces of equal size but **in opposite directions**, the forces will balance each other out.

Newton realized that a thing will only change its speed or direction of travel when pulled by a force which is not balanced by other forces.

This is a very important idea. In everyday life we often think that a force is needed just to keep something moving at the same speed. In fact this force is needed to overcome friction. Friction is a force which acts against motion (see page 45).

The best way to understand Newton's first law is to think of a spaceship moving in outer space so far from any star or planet that there is no gravity.

The spaceship has no engine to push it but there is no friction or other force to slow it down. It just travels on and on at the same speed and in the same direction.

The spaceship could be speeded up or slowed down by firing little rockets fixed to it. These would give a force to change the spaceship's speed or direction.

Things which are not moving have a speed of 0 m/s. These stationary objects obey Newton's first law too. Quite simply, as long as there are no unbalanced forces acting on the object, it will not start to move.

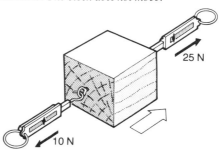

These forces balance. The block does not move.

These forces do not balance. The block moves as if it was being pulled by a force of 15 N to the right.

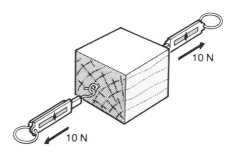

Space capsule; obeying Newton's laws

Activities

Try to find out about Sir Isaac Newton.
- **a)** When was he born? When did he die?
- **b)** Where was he at university?
- **c)** What was the name of the book which contained his laws of motion? When was it published?
- **d)** What other scientific work did he do?
- **e)** What did he have to do with
- i) the Houses of Parliament
- ii) the Royal Mint?

Acceleration

When an unbalanced force acts on a moving object in the direction of its travel, the object's speed changes. The force may make it speed up but if it is in the opposite direction it will slow the object down.

Acceleration tells us how fast the speed of something is changing.

Starting from rest, each of the cars (shown opposite) can get up to a speed of 60 + miles per hour (30 m/s). Car A takes 10 s to reach 60 + mph. Car B can reach 60 + mph in just 6 s! Car B has the greater acceleration.

Acceleration can be shown very clearly on a speed–time graph. The one shown opposite has two lines; one for car A and one for car B.

The line for the sports car (B) is steeper than the line for car A. **A steep slope on a speed–time graph means a large acceleration.**

We calculate the acceleration by working out how much the speed changes in one second. For example, a racing car starts from rest. After 6 s it has accelerated to a speed of 30 m/s. Its speed has changed by 30 m/s in 6 s. In one second its speed would have changed by $30/6 = 5$ (m/s)/s.

We can say the units as 'metres per second, per second'. This is because the acceleration tells us how much the speed has changed (in metres per second) in each second. [The units are often written as m/s^2.]

$$\text{acceleration} = \frac{\text{change in speed}}{\text{time taken}}$$

Worked example

1 A sprinter starts from rest (0 m/s) and takes four seconds to get up to a speed of 12 m/s. What is the acceleration?

The change of speed is $12\,\text{m/s} - 0\,\text{m/s} = 12\,\text{m/s}$. The time taken is 4 s.

$$\text{acceleration} = \frac{12\,\text{m/s}}{4\,\text{s}} = 3\,\text{m/s}^2$$

Ready, steady, go!

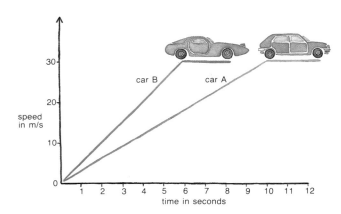

Accelerations from speed–time graphs

We have already seen that the 'slope' of a line on a speed–time graph tells us about acceleration. To work out a value for this we just divide the distance that the line rises (change of speed) by the time taken. The graph below shows how.

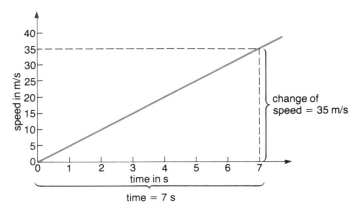

(*Note:* you must remember to take the numbers from the axes on the graph. It does not work if you 'count squares' or measure with a ruler.)

Slowing things down

Friction

Our everyday experiences of motion do not seem to fit Newton's first law. We know that if we stop pedalling our bicycle it will slow down and stop. It is as if we need to keep pushing with a force just to keep a steady speed! However, this force is needed to balance other forces which act against motion. One of these is called **friction**.

Friction always appears when we try to slide one thing over another. This is because all surfaces, even polished ones, are rough when seen under a microscope. If we try to push one surface over another, the rough pieces catch on each other. This causes a force against our push.

When we want things to keep moving, friction is a nuisance because we have to use energy to overcome it. **Lubricants** such as oil can make things move more easily. When we want things to stop we can use friction. When bicycle brake blocks rub on the rim of the wheel, the friction slows the bicycle down.

Surface of a piece of paper greatly magnified

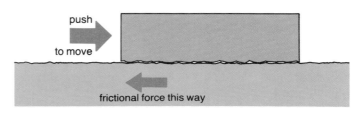
Sliding surfaces

Air resistance

When cars, bicycles and other things try to move in air, there is a force against them called **air resistance**. This is because air is made up of gases. The molecules in these gases bump into moving objects causing a force against the motion. At low speeds this is only a small force. At higher speeds it becomes much larger. Air resistance can be reduced by using shapes which let the air 'slip' past more easily. This is called **streamlining**.

Now that we have met friction and air resistance we can see how Newton's first law applies.

Bicycle brakes; using friction

Accelerating

Steady speed

Decelerating

To accelerate, the force from the engine must be bigger than the forces of friction and air resistance put together.

At steady speed, the forward force will just balance the forces of friction and air resistance.

To stop a car the driver takes his foot off the accelerator pedal. This reduces the force from the engine. He also puts on the brakes. This makes the friction much larger.

Stopping a car

A car driver needs to be able to stop in a short distance to avoid accidents. To stop the car, the driver presses on the brake pedal. This makes the brake pads or brake 'shoes' rub against the wheels. The friction makes the car slow down but it takes some time before the car stops completely.

To find the stopping distances of a car a simple test can be carried out. The driver drives the car at a steady speed. At a given signal, the brakes are applied to bring the car to rest as quickly as possible without losing control. This is sometimes called an emergency stop.

The speed–time graph opposite shows what happened in a test where the car was moving at 12 m/s (about 30 mph).

The first thing to notice is that the car did not start slowing down until 0.75 s after the signal. This is how long it took the driver to react. It is sometimes called '**thinking time**'. While the driver was thinking, the car travelled 9 m!

Even with good brakes the car takes about 2.5 s to stop. During this time it travels 15 m more.

You can see that even at a fairly low speed of about 30 mph, a car takes about 24 m to stop. If the car moves at higher speeds it will take a much longer distance to stop. At 30 m/s (about 70 mph) the total stopping distance is 112.5 m; longer than a football pitch!

Questions

1 Alcohol in the blood stream slows down a person's reactions. Why is a driver who has been drinking alcohol more likely to have an accident?

2 Cars have lights at the back which come on as soon as the driver presses the brake pedal. Why is this important? Suggest what would happen if a car had brake lights which did not work.

3 A driver is travelling at 24 m/s (about 55 mph) on a motorway in fog when he sees that there has been an accident 50 m in front. He takes 0.5 s to react before pressing on the brake pedal. The brakes take 4 s to stop the car.

    a) Draw a speed–time graph to show what happens. Your time axis needs to go from 0 to 5 s.

    b) Work out the total stopping distance.

    c) Comment on your answer.

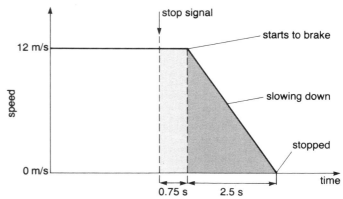

Thinking time = 0.75 s
Distance travelled during thinking time = 12 m/s × 0.75 s = 9 m
Distance travelled during braking = ½ × 2.5 × 12 = 15 m
Total stopping distance = 9 m + 15 m = 24 m

Activities

Measuring reaction times

The apparatus shown below can be used for measuring reaction times. It uses an electronic clock which measures times as short as 0.01 s ($\frac{1}{100}$th of a second). Not all electronic clocks have the same 'start' and 'stop' controls so your circuit may be slightly different; the idea will still be the same.

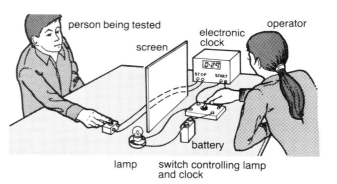

Testing reaction time

The operator's switch does two things; it turns on a signal lamp and, **at the same time**, it starts the clock. The other switch stops the clock.

When the lamp comes on, the person being tested must press his or her switch as quickly as possible.

When measuring someone's reaction time, you should let them have two or three trial runs so that they get used to the apparatus. Then record their reaction time for about five trials. Work out the average reaction time.

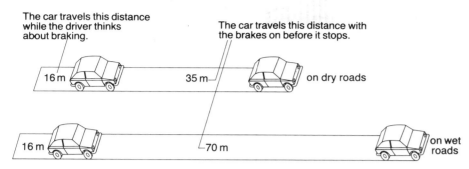

Stopping distances

Friction and road safety

We could not drive a car without friction between the tyres and the road. Three reasons are given below.

When the engine turns the wheels, friction lets the car move forward. When the roads are icy you may see cars with their wheels slipping, unable to move.

When the driver turns the steering wheel, friction lets the car change direction. When the roads are wet or icy, cars may spin off the road at sharp bends. Friction has not been big enough to keep the car on the road.

When the driver puts the brakes on, friction between the tyres and the road lets the car slow down. On wet roads the water acts like a lubricant. This makes braking more difficult so drivers should take more care.

Wet roads

If large puddles form on a road when it rains, a car travelling at high speed may 'ski' on top of the water. The driver cannot brake or steer! To prevent this, roads are built with a slight curve called a **camber**. The camber makes the rain run to the side of the road into drains. Even with this, a thin layer of water is left on the road. The tread of a car tyre has patterned grooves to let the water escape from under the wheels.

Dry roads

Scientists have developed special 'high-friction' materials to use as road surfaces. These materials allow cars to stop quickly without skidding. However, they are more expensive than ordinary 'tarmac' so they are only used where accidents are more likely. For example, they could be used for a 50 m length of road near to a pedestrian crossing.

Icy roads

When the roads are covered in ice driving is very dangerous. To prevent accidents, salt and grit are spread on the roads. The salt helps to melt the ice and the grit helps to increase friction.

Skidding

When a car is travelling on wet or icy roads, it is easy to brake so quickly that the wheels 'lock'. This means that they suddenly stop turning but the car skids on with only friction to stop it. The driver loses control.

Police drivers are taught to quickly release the brake pedal and then to press it again as soon as the wheels start turning. This press-release-press-release method is called **cadence braking** and stops the car in a much shorter distance.

The car shown below has a braking system which is controlled by an electronic 'brain'. If the brake pedal is pushed so hard that the wheels lock, the electronic control system releases the brakes and then applies them again. It can do this many times in one second . . . better than any human driver.

Car with advanced braking system (ABS)

Activities

1 Find out about 'snow chains'. How do they work?

2 Find out about tyres used by racing cars. Do they always have patterned tread?

3 Watch a Grand Prix race on television. What happens at a pit stop? Why?

Forces and acceleration

Unbalanced forces cause changes in speed. As scientists, we want to know the size of the acceleration given by a force. This helps us to understand movement and helps us to design cars, aeroplanes and other machines.

Common sense tells us part of the answer. The athlete shown opposite knows that if she can apply a bigger force then the javelin will go further. This is because bigger forces will accelerate it to larger speeds.

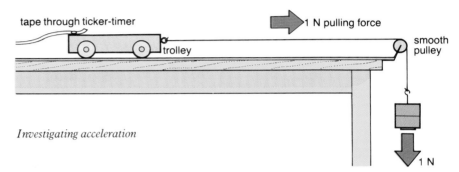

Investigating acceleration

More force means more acceleration

We can find out more about the connection between force and acceleration by observing what happens to a trolley when it is pulled by a known force. The diagram above shows one method. The weight hanging on the string, pulls the trolley with a force of 1 N. When we let go of the trolley it accelerates, pulling the tape through the ticker-timer. We can cut up the tape to make a speed–time graph.

To find out what happens when the pulling force doubles we just add another 1 N weight to the string. The experiment is repeated and the tape cut up to make a new speed–time graph.

The graphs opposite show that the slope of the speed–time line is steeper when the force is 2 N. In fact the acceleration is twice as large.

After doing the experiment several times with different forces we find that:

twice the force gives twice the acceleration,

three times the force gives three times the acceleration,

four times the force gives four times the acceleration,

and so on.

We can say that **the acceleration is proportional to the force applied**.

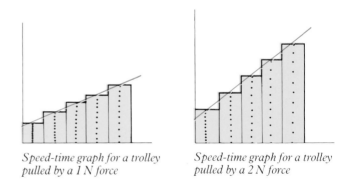

Speed-time graph for a trolley pulled by a 1 N force

Speed-time graph for a trolley pulled by a 2 N force

Worked example

A car breaks down. The driver's friend pushes it with a force of 200 N. It accelerates at 0.25 m/s^2.

What would the acceleration be if a second friend pushed with another 200 N?

A force of 200 N gives an acceleration of 0.25 m/s^2.

When the second friend pushes the force is
200 N + 200 N = 400 N.

The force has doubled so the acceleration must have doubled.

$$0.25 \text{ m/s}^2 \times 2 = \textbf{0.5 m/s}^2$$

Mass and acceleration

What is mass?

When we want to know how much material an object contains we measure its **mass**. The mass is measured in kilograms (kg).

The mass of an object tells us something about how hard it will be to move. To test this, hang an empty can from a piece of string fixed to the ceiling. Next to it hang an identical can filled with sand. The empty can will have a much smaller mass than the one filled with sand. Now try giving each can a small, sharp push. Which one moves more easily? Set the cans swinging. Which one is the easiest to stop?

The bag of sugar has a mass of 1 kg. The cement has a mass of 50 kg. The bag of sugar is 'easier' to move.

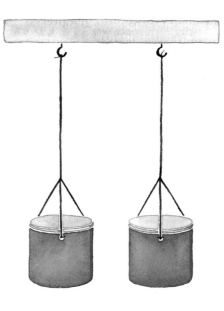

Which one is filled with sand?

To see how the mass affects acceleration we can load a trolley with extra masses and use a ticker-timer to investigate the motion as it is pulled by a steady force.

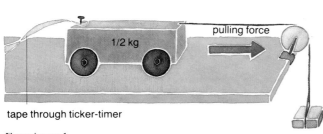

Experiment 1

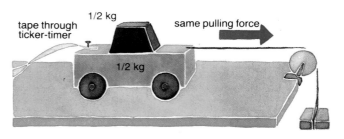

Experiment 2

Not surprisingly we find that the force produces **less acceleration** when the trolley carries **more mass**. We find that the slope of the speed–time line for the 1 kg trolley is only half that for the $\frac{1}{2}$ kg trolley.

If the mass is doubled the acceleration is halved.

If the mass is trebled the acceleration is only one third ($\frac{1}{3}$) as large and so on.

We can say that **the acceleration is proportional to $\frac{1}{mass}$**.

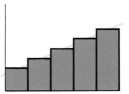

Speed-time graph for force pulling ½ kg (Experiment 1)

Speed-time graph for force pulling 1 kg (Experiment 2)

Questions

1   A good racing car will give large accelerations. Explain why racing car designers use low density materials like fibre glass for the body work.

2   Why do racing cars have powerful engines?

Your answers should include the words 'force', 'mass' and 'acceleration'.

A pattern for force, mass, and acceleration

We have seen that the acceleration of a car, a tennis ball, or any other object depends on two things; mass and force. Newton's second law of motion puts these things together.

You should remember that forces are measured in units called newtons. One newton is defined so that the relationship between force, mass and acceleration is very simple.

DEFINITION

1 N force acting on a mass of 1 kg causes an acceleration of 1 m/s^2.

Using this definition and Newton's second law we can build up a table of results.

unbalanced force F	mass m	acceleration a	comment
1 N	1 kg	1 m/s^2	definition of 1 N
2 N	1 kg	2 m/s^2	The force has doubled so the acceleration has doubled.
3 N	1 kg	3 m/s^2	Now the force is 3 N so the acceleration is three times as large.
1 N	2 kg	$\frac{1}{2}$ m/s^2	The force is 1 N but the mass has doubled. The acceleration has halved.
2 N	2 kg	1 m/s^2	Here the force is 2 N but it acts on 2 kg. The extra force and extra mass 'cancel out' so the acceleration is 1 m/s^2.
6 N	3 kg	2 m/s^2	The force is 6 N but it is acting on a mass of 3 kg. The acceleration is $\frac{6}{3} = 2$ m/s^2.

We can see from the table that the number in the 'force' column is always equal to the numbers in the other two columns multiplied together.

We can write: **force = mass × acceleration**,

or in symbols: $F = ma$

Warning! You should always use newtons, kilograms and m/s^2 in this equation.

Worked examples

1 The car shown in the diagram has a mass of 600 kg. The man is pushing with a force of 500 N but there is a frictional force of 200 N. What is the acceleration of the car?

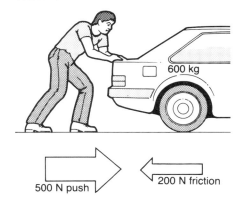

500 N push   200 N friction

STEP 1 The unbalanced force
= man's force − friction

= 500 N − 200 N

= 300 N

STEP 2 Use $F = ma$
$F = 300$ N and $m = 600$ kg

$$F = m \times a$$

$$300 \, \text{N} = 600 \, \text{kg} \times a$$

$$a = \frac{300 \, \text{N}}{600 \, \text{kg}} = \frac{1}{2} \, \text{m/s}^2$$

The car accelerates at 0.5 m/s^2.

2 A spacecraft has a mass of 500 kg. It is in outer space where there is no air resistance. It fires its rocket motor and accelerates at 2 m/s^2. What force does the rocket motor give?

Use $F = ma$ $m = 500$ kg
and $a = 2$ m/s^2

$$F = m \times a$$

$$F = 500 \, \text{kg} \times 2 \, \text{m/s}^2$$

$$F = 1000 \, \text{N}$$

The engine gives a thrust of 1000 N.

Energy and motion

Forms of energy

When you make a thing move by pushing it or lifting it up you are doing work and using energy. Machines also use energy to do work.

We can think of several types of energy.

- **Chemical energy** is stored in fuels. Fuels like petrol release energy when they burn.
- **Electrical energy** can be produced by a battery or a dynamo. In the home it is changed to light (and some heat) in lamps and heat in cookers.
- **Radiation energy** is given out by hot objects and travels as waves. Radiation energy from the sun heats the surface of the Earth.
- **Sound energy** travels as a wave from vibrating objects such as loudspeakers and musical instruments.
- When a spring or piece of elastic is stretched it has **potential energy**. When released it will spring back giving out the stored energy. Similarly, when you pick up a book from the floor it has potential energy. If you release it, it falls down giving out the stored energy. The energy of an object which has been lifted up is called **gravitational potential energy**.

The potential energy stored in the stretched catapult elastic is changed to kinetic energy of the moving stone.

The potential energy of the skier at the top of the slope is changed to kinetic energy.

All forms of energy can be used to make things move. When this happens they are converted into another form called kinetic energy. **All moving objects have kinetic energy**.

We can use the kinetic energy of an object to do work. When a car stops its kinetic energy changes to other forms. Some of it will appear as heat and the brakes will get hotter. In large vehicles like trains the brakes may have to be cooled with water to stop them becoming too hot.

Working out how much kinetic energy

Energy is measured in units called **joules**. You need one joule of energy to move a force of 1 N though a distance of 1 m.

When you lift a 1 kg bag of sugar from the floor on to a table 1 m high you need a force of 10 N to lift it.

work done = force × distance

so the work you do is 10 N × 1 m = 10 Nm. This needs 10 J of energy.

If the bag falls off the table, its gravitational potential energy (10 J) will change to kinetic energy.

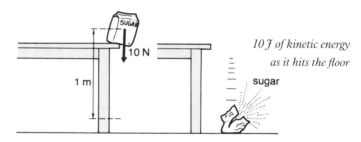

10 J of gravitational potential energy

The kinetic energy of a moving object can be calculated from the following formula:

$$\text{K.E.} = \frac{1}{2}mv^2 \quad (m = \text{mass of object}$$
$$v = \text{speed} \;\; v^2 = v \times v).$$

Worked example

A car of mass 500 kg is moving with a speed of 10 m/s. What is its kinetic energy?

$$\text{Kinetic energy} = \tfrac{1}{2}mv^2$$
$$= \tfrac{1}{2} \times 500\,\text{kg} \times 10\,\text{m/s} \times 10\,\text{m/s}$$
$$= 25\,000\,\text{J}$$

The car has 25 000 J of kinetic energy.

Questions

1 Work out the kinetic energy of the following.
   a) A train of mass 100 000 kg travelling at 10 m/s.
   b) A runner of mass 60 kg running at 5 m/s.
   c) A skier of mass 50 kg moving at 30 m/s.

Power

Engines and work

Engines allow machines to do work. Some engines turn wheels to make vehicles move, others lift weights or pull things along. They are using energy to make forces move.

For example a crane uses an engine to turn a drum which winds up a cable. This lifts a weight. As the weight rises it gains gravitational potential energy. The power of the crane tells us how quickly it can change the potential energy of the load being lifted.

The power of an engine tells us how quickly it can convert energy.

$$\text{power} = \frac{\text{energy}}{\text{time}}$$

The unit of power is the watt (symbol W). One watt is equal to one joule of energy in each second.

The example below shows how to calculate power using the equation.

Power of a crane

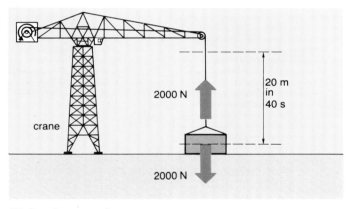

Finding the power of a crane

A crane can lift a load of 2000 N up through a height of 20 m in 40 s.

Now it takes 1 J of energy to lift 1 N through 1 m.

The crane converts 2000 N × 20 m = 40 000 J in lifting the weight.

This takes 40 s so the power is 40 000 J/40 s = 1000 J/s.

The power output of the crane is 1000 W or 1 kW.

(1000 W = 1 kilowatt = 1 kW.)

Brake horse power

Car engines often have their power given as **brake horse power** (bhp). This is the power of the engine when working against a brake, but measured in units called 'horse power'. This dates back to the days when it was defined by measuring the work which a cart horse could do! For some strange reason it is still used for engines.

One horse power is approximately equal to 750 W.

Doing work against a brake!

Questions

1 Look at these performance figures for new cars. They were taken from a car magazine.

model	top speed	time to go from 0–60 mph	engine power
Mini	82 mph	17.2 s	40 bhp
Ford Escort (1.1)	98 mph	13.5 s	60 bhp
Jaguar XJ6 (2.9)	112 mph	10.0 s	165 bhp
Porsche 911	152 mph	6.1 s	231 bhp
Porsche 911 (turbo charged)	171 mph	5.1 s	330 bhp

a) Which column in the table tells you about acceleration? Explain your answer.
b) Which car has got the smallest acceleration?
c) Which car has got the most powerful engine?
d) What do you notice about top speed and engine power?
e) Try to find out about turbo chargers.

Measuring your own power

The human body uses chemical energy from food to do work. We do work when we use a force to make something move. For example, when we push a wheelbarrow, drag a heavy sack across the floor or lift a weight, we are doing work. To find the amount of work done we need to measure the size of the force used and the distance it moved. The human body does work like a machine and so we can work out its power.

If the sack moves 20 m the girl has done 120 N × 20 m = 2400 Nm of work. The energy needed to do this is 2400 J.

The activity below will give you a rough estimate of your body's power.

Activities

Stage 1

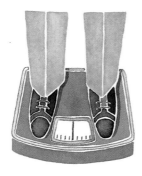

Find out how much force you will need to lift your body by weighing yourself on scales. Most bathroom scales give you your mass in kilograms. To convert this to newtons just multiply by 10. For example, if you have a mass of 35 kg your weight is 350 N.

This is because 1 kg weighs about 10 N on the Earth's surface.

Stage 3

Get someone to time you with a stopwatch as you run up the stairs as fast as you can. If you want, try it a few times and take your fastest time.

Stage 2

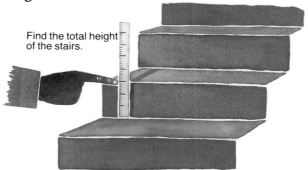

Find the total height of the stairs.

You are going to run up a flight of stairs. Find out how high you will raise your body weight by measuring the height of the stairs in metres. Do this by measuring the height of one step and then multiplying by the number of steps. For example, 15 steps each 20 cm high will give a height of 3 m (300 cm).

Stage 4

Work out your power. (A calculator will help!)

$$\text{energy used} = \text{weight} \times \text{height of stairs}$$

$$\text{power} = \frac{\text{weight} \times \text{height of stairs}}{\text{time taken}}$$

For example, Sarah weighs 400 N. She runs up stairs 3 m high in 5 s.

Her power is $\dfrac{400\,\text{N} \times 3\,\text{m}}{5\,\text{s}} = 240\,\text{W}$

Sarah's power output is 240 watts; enough to light several light bulbs!

Momentum

Commonsense tells us that fast moving things are more difficult to stop than slow moving things. Think about stopping a cricket or hockey ball which is rolling slowly towards you. Then think about stopping the same ball when it has been hit towards you at high speed! We say that the faster ball has more **momentum**.

Momentum also depends on mass

The lorry shown opposite is moving at 30 m/s. To stop it the driver must apply a force. Because the lorry has a very large mass the brakes will need to give a large force. If a smaller force is used it will take longer to stop the lorry. The car shown is also moving at 30 m/s. We can use a much smaller braking force to stop this in the same time as the lorry.

The lorry has got more momentum because it has got more mass.

Momentum is given by mass × velocity

Its units are kg m/s (kilogram metres per second).

A force is always needed to change the momentum of an object. This is true whether we are speeding things up or slowing them down. The size of the force and the time it acts determines how much the momentum will change.

The diagrams opposite show that if we use the same force for the same time we always get the same change in momentum. A bigger force would have caused a bigger change. The momentum would have also changed more if we had applied the force for more than 1 s.

force × time = change of momentum

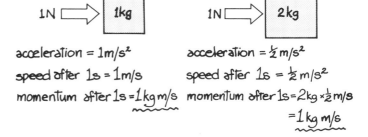

$1N \Rightarrow$ 1kg $1N \Rightarrow$ 2kg

acceleration = 1m/s²   acceleration = ½ m/s²

speed after 1s = 1m/s   speed after 1s = ½ m/s²

momentum after 1s = 1kg m/s   momentum after 1s = 2kg × ½ m/s = 1 kg m/s

Questions

1 Calculate the momentum of
   a) a 1000 kg car moving at 5 m/s
   b) a 10 000 kg lorry moving at 5 m/s
   c) a 10 000 kg lorry moving at 20 m/s
   d) a 0.1 kg ball moving at 10 m/s.

2 An American footballer has a mass of 110 kg. He can run at 10 m/s. What is his momentum? Why is he difficult to stop?

Worked example

A lorry of mass 10 000 kg is moving at 10 m/s when it starts to brake. The braking force is 4000 N. How long does it take to stop?

change of momentum = (10 000 kg × 10 m/s) moving − (10 000 kg × 0 m/s) stopped

change of momentum = 100 000 kg m/s

force × time = 100 000 kg m/s

time = $\frac{100\,000 \text{ kg m/s}}{4000 \text{ N}}$ = 25 s

The lorry takes 25 seconds to stop.

More momentum

Newton's first law tells us that without an unbalanced force, things just keep moving in the same direction with the same speed. You may have experienced one result of this law if you have been standing in a bus when it has started off. You feel like you have been thrown backwards but in fact you have been left behind because you have not accelerated at the same rate as the bus.

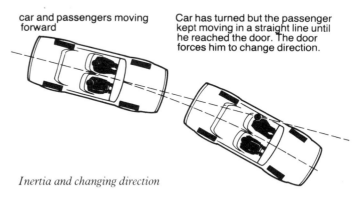

car and passengers moving forward

Car has turned but the passenger kept moving in a straight line until he reached the door. The door forces him to change direction.

Inertia and changing direction

The bus starts to move. The passenger is 'left behind' because she has not accelerated as fast as the bus.

Similarly you may have been in a car which has turned a corner too quickly and have been 'thrown' against the door. What has happened is that the car has changed direction but you have kept going in a straight line! This is because the friction between you and the seat has not been big enough to change your direction quickly enough.

This tendency to keep going in the same direction and at the same speed is called **inertia**. It can have serious consequences if the vehicle's motion changes very suddenly, for example in a crash.

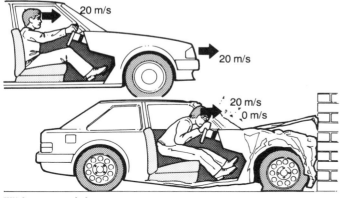

20 m/s

20 m/s

20 m/s
0 m/s

Without a seat belt

Activities

1 Find out why airlines make you wear a seat belt on landing **and** on take-off.

2 Find out about inertia reel seat belts.

3 It is illegal for drivers not to wear seat belts. Ask people who have been driving for several years whether they would wear a seat belt if it were not compulsory. What percentage would?

force on driver due to seat belt

With a seat belt

The diagrams show what happens to a passenger in a car moving at 20 m/s when it stops suddenly. When the car hits the wall its speed suddenly becomes 0 m/s. The passenger keeps moving at 20 m/s until he hits the windscreen.

To stop this sort of thing happening all passengers in a car should wear seat belts. Then in an accident the seat belt will provide a force to slow the passenger down.

Safety in crashes

We have seen that seat belts will stop passengers in a car from travelling forward in a crash but what happens if a car is hit from behind?

If a parked car is hit from behind, the impact will make it accelerate forwards. The passenger is also accelerated because the back of the seat pushes against her.

However, her head is above the back of the seat and, because of its inertia, it does not move forward so quickly. It feels as if it has been thrown backwards.

The sudden movement can damage the bones of the neck. It can also damage the nerves connected to the spine. This can cause pain in the neck and arms. In serious cases it can cause temporary paralysis of the arms and shoulders. Injuries caused in this way are called '**whiplash' injuries**.

To prevent 'whiplash' injuries some cars are fitted with head restraints. These stop the head from going too far back and damaging the spine.

Safety cages

In an accident large forces will act on the car. It is important that these do not crush the part of the car where the people sit. Car designers make sure that this part of the car is built of strong steel tubes welded together to give a protective cage.

reinforced 'cage' for passengers

'Crumple zones'

It is very important for the safety cage to be strong and rigid. Perhaps surprisingly we want the rest of the car body to bend and distort in an accident. This will let the impact take a longer time. The car comes to a stop more slowly and so the forces are not so big. Inside the car the steering column should be collapsible.

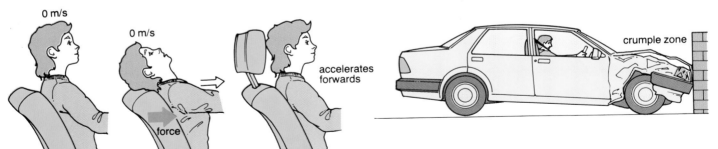

0 m/s   0 m/s   accelerates forwards

force

before impact   during impact

crumple zone

Activities

1 Try to get some information from motor manufacturers on their latest models. What safety features do they have?

2 Carry out a survey in a car park. (If it is a public car park check with the attendant first!)
Find out what fraction of cars have seat belts or child seats in the back. What fraction of cars have head restraints in the front and in the back? The table below shows one way of collecting data.

Rally car with extra protection

Make	Model	Year	Rear seat belts	Front head restraints	Rear head restraints
Ford	Granada	E-reg. (1988)	✓	✓	✓
Ford	Escort	T-reg. (1979)	✗	✗	✗

Questions

1 Oil is a lubricant. This means that it can reduce the friction between moving parts.

    a) Why is it good to oil the axle of a bicycle wheel?

    b) Why would it be dangerous to put oil on the rim of a bicycle wheel?

2 The car shown below is designed to travel at very high speeds. To stop it uses ordinary brakes and a parachute.

How does the parachute help the car to stop quickly?

Drag racer

3 The figure below shows the speed–time graph of a drag racer (a car designed to race over short distances).

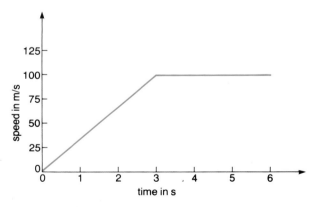

Speed-time graph for a dragster

a) How fast was the drag racer going after
i) 1 s ii) 2 s iii) 5 s?

b) What was the acceleration of the drag racer?

c) Roughly how long did the drag racer take to reach 60 mph (27 m/s)?

d) How far had the drag racer travelled in 6 s?

e) An ordinary car has an engine power of about 90 bhp. The drag racer in this question had an engine which gave 1700 bhp. Why?

4 The downhill skier shown has wax on the bottom of his/her skies. Suggest why this helps him/her to travel faster.

Downhill skier

What things can you see in the photograph which help to reduce air resistance? (*Hint*: there are at least three things.)

5 List the energy changes which are taking place in this story.

'After eating her breakfast, Sarah got on her bicycle and rode to school. She got up to quite a good speed and was feeling quite warm by the time she reached the hill at the end of the road. She managed to cycle up the hill without stopping but had to rest at the top for a few minutes. She then let the bicycle free-wheel down the hill. Even without pedalling the bicycle accelerated so much that at the bottom of the hill Sarah had to brake hard in order to stop the bicycle before she crashed into the school gates.'

6 The poster shown below was designed to persuade people not to drink and drive.

    a) Explain why it is dangerous to drive after drinking alcohol.

    b) Why is this even more dangerous on a wet night?

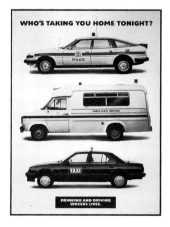

Do posters like this work?

What is inside our planet?
How do different types of rock form?
What causes earthquakes?
Are our continents moving?

EARTHQUAKE ROCKS MEXICO CITY

ARMENIA DEVASTATED IN USSR'S WORST EARTHQUAKE

Chilean earthquake registers 8.3 on the Richter Scale - releases estimated 10^{13} MJ of energy

Krakatoa ERUPTS - 36,000 killed

If you live in the United Kingdom you probably think that our planet is hard, solid and safe. The headlines on this page show that in other parts of the world there is a lot of evidence to show that this is not the case. The Earth's crust is constantly moving and can sometimes react violently. By carefully studying vibrations and sudden movements in the Earth's crust, scientists can sometimes predict a big earthquake or the eruption of a volcano. This can help to save hundreds of lives.

The processes inside the Earth that cause volcanoes also form all of our rocks and minerals. We use these for building materials, extract metals from them, and use them as fuels and raw materials for our industries. It is important to study the internal structure of the Earth so that we can make the most of our resources.

Inside the Earth

Scientists who study the rocks are called **geologists**.

They get information about the inside of the Earth in four main ways: by studying the rocks on the surface of the planet, by studying the materials ejected from volcanoes, by recording vibrations inside the Earth and by drilling far down into the crust and taking out rock samples.

Using all of these methods we have built up a good picture of the inside of our planet.

The Earth is thought to be made up of three main layers:

- the **core** has two parts: a solid inner core consisting of a mixture of nickel and iron, and an outer core made of molten iron with some other elements. The temperature in the core is between 3700°C and 4500°C!

- the **mantle** is a little more complex. Near the core the rocks are made mainly of iron and magnesium silicates. Nearer to the surface the mantle is composed of igneous rocks, mainly peridotite. Mantle temperatures are between 1000°C to 3700°C.

- the **crust** is the thin layer of rocks directly under our feet. It contains rocks of lower density such as basalt and granite. Temperatures are very much cooler than those found deeper inside the Earth.

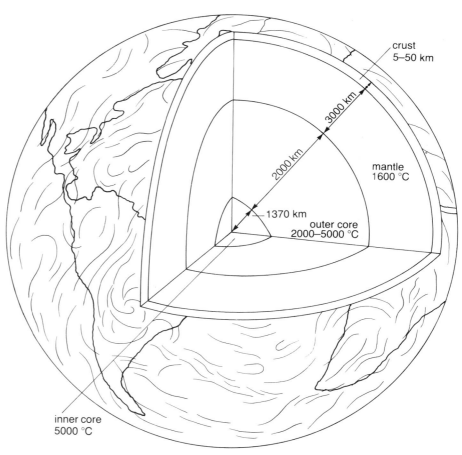

The structure of the Earth

Questions

1 Give **two** reasons for believing that it is hot inside our planet.

2 Give **two** reasons why geologists study volcanoes.

3 What two metals would you find at the centre of the Earth?

4 Explain why the word 'crust' is a good description of the layer of rocks near to the Earth's surface.

5 The deepest hole ever drilled is about 12 000 m (12 km) deep. The temperature at this depth is about 200°C.

a) Give **one** reason why geologists drill such deep holes.

b) Suggest how we could make use of the hot rocks deep in the Earth's crust.

6 Imagine you are making a journey to the centre of the Earth. Describe the conditions that you would find as you travel from the surface to the inner core.

How old is the Earth?

For many generations people have tried to work out the age of the Earth. Parts of its surface are extremely old – up to 4500 million years! Try to imagine what that means.

To help us to appreciate the enormous time scales involved in the evolution of the Earth, scientists have constructed the chart shown opposite. This chart is often referred to as the **stratigraphic column** for the Earth. It is made by drawing blocks to represent rocks of known ages on top of one another. The oldest are at the bottom and the youngest are at the top.

The whole of Earth history is divided up into four major divisions called **eras**. Each era is further subdivided into **periods** or epochs. Rocks formed in these periods are named accordingly, for example, rocks between about 500 and 570 million years old belong to the Cambrian period.

Measuring the age of rocks

Fossils can help determine whether one rock is younger than another, but they can tell us nothing about the age of a particular rock.

Fortunately rocks that contain radioactive elements have a built-in clock for measuring their age. Elements such as uranium are unstable and slowly break down to form other elements. Uranium breaks down into lead. The period of time needed for half the nuclei in the element to break down is constant and is called the **half-life** (see page 101). If the amount of uranium and lead in a rock are measured then its age can be calculated.

Chapter 5 will tell you more about **radiometric dating.**

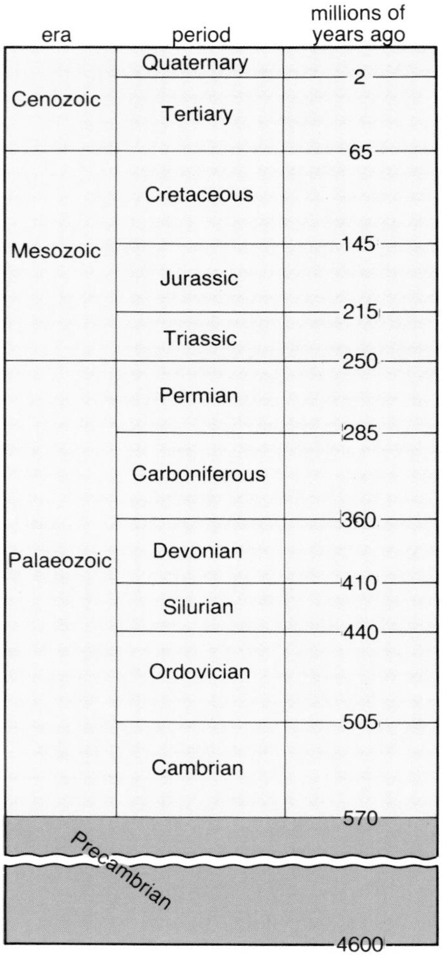

era	period	millions of years ago
Cenozoic	Quaternary	2
	Tertiary	
		65
Mesozoic	Cretaceous	
		145
	Jurassic	
		215
	Triassic	
		250
Palaeozoic	Permian	
		285
	Carboniferous	
		360
	Devonian	
		410
	Silurian	
		440
	Ordovician	
		505
	Cambrian	
		570
Precambrian		
		4600

The geological eras

Activities

1   Copy and complete the table below using the information from the stratigraphic column.

rock	location	age (million yrs)	era	period
Granite	Peterhead, Aberdeenshire	385		
Basalt	Auvergne, France	10		
Peridotite	Ivrea, Italy	70		
Sandstone	Leeds, England	300		
Limestone	Derbyshire, England	320		

2   Use a geological map in an atlas to find the age of the rocks near your home. What period do they date from?

Questions

1   Roughly how old is the Earth?

2   Fossils of fish are found in the Ordovician period, but not in earlier rocks. Roughly how long ago did fish appear on the Earth?

3   Fossils of dinosaurs are found in Cretaceous rocks but not in Tertiary rocks. Roughly how long ago did dinosaurs become extinct?

4   Coal deposits are from the Carboniferous period. Roughly how old is a lump of coal?

5   Explain why rocks containing radioactive elements can be dated more accurately than those that do not (see chapter 5).

The rock cycle: sedimentary rocks

Once igneous rocks are exposed to the atmosphere on the surface of the Earth, they start to be broken down. The silicate minerals of the igneous rock are attacked by chemical and physical processes that lead to their destruction. This is known as **weathering**. The products of the weathering action accumulate to form **sedimentary rocks**. Sandstone, clay, limestone and chalk are sedimentary rocks.

There are two main types of weathering:

Physical weathering

Rocks can be worn away or **eroded** by the action of wind but perhaps the best example of physical weathering is frost shattering. Water seeps into cracks in the rock and then freezes. As it turns to ice it expands and forces the rock to split. Large boulders fall and shatter to form **scree** at the foot of mountain slopes. You can see the effect of frost action on road surfaces after a severe winter.

Scree

Chemical weathering

Scree fragments and exposed rock faces are open to attack by rain-water. Rain-water absorbs carbon dioxide from the air making it slightly acidic. Chemical reactions take place between the water and the rock minerals causing them to crumble and possibly be dissolved and carried away.

Some minerals such as quartz are not affected by chemical weathering whereas others like feldspar are broken down to clay minerals. In Cornwall the minerals in the granite have been broken down to form china clay – the basis of the North Staffordshire pottery industry.

Transporting the weathered material

Weathered rock fragments are washed into rivers.

Rock fragments are deposited as sand and gravel in the sea.

Rivers carry rock fragments to the sea.

Formation of sedimentary rocks

Over millions of years sediment will pile up forcing water out of the lower layers and compacting them together, eventually forming rock. The size and origin of the sediment particles will determine the type of sedimentary rock that is produced. Sandstone for example is formed from compacted sand grains. Limestone contains a large proportion of organic material such as calcium carbonate from the shells and skeletons of marine creatures.

Questions

1 Explain why a cold winter can cause damage to road surfaces.

2 Describe how a mountain is eroded.

3 What is sediment? Why do the mouths of rivers sometimes need to be cleared by dredgers?

4 Beach sand contains very small, smooth sand grains of different colours. How do you explain these facts?

The rock cycle: metamorphic rocks

Metamorphism means 'change of form'. Many sedimentary and igneous rocks have been subjected to mechanical forces, squeezing and pushing, and to different conditions of temperature and pressure since they were formed. These processes produce a new group of rocks called **metamorphic rocks**.

For example, rock can be changed when very hot magma is pushed up into it. Where the magma pushes into other rocks, it is called an **intrusion**. The size of the intrusion is important. Small intrusions will not heat up the surrounding rock as much as large ones. The diagram shows how the metamorphic rocks marble and quartzite are made as a result of a magma intrusion into bands of sedimentary limestone and sandstone. Rather like a cake being baked, the end product is very different to the original mixture.

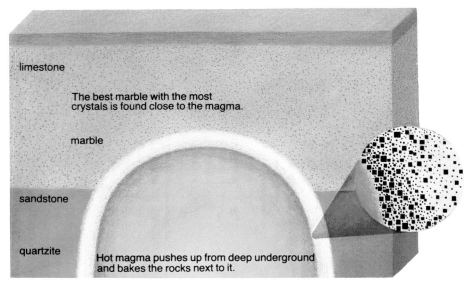

limestone

The best marble with the most crystals is found close to the magma.

marble

sandstone

quartzite

Hot magma pushes up from deep underground and bakes the rocks next to it.

Making metamorphic rocks

Metamorphic rocks are extemely hard (like burnt cakes!) and usually have a spotty appearance due to the formation of new minerals inside them. The type of rock formed depends upon the area of contact between the magma intrusion and the surrounding rock. The best marble is formed immediately next to the magma.

Splitting slate

Regional metamorphism affects very large masses of rock and is usually associated with mountain building. Most of the continents of the Earth are regionally metamorphosed rocks. Huge movements of the Earth subjected rocks to great pressure and high temperatures. This formed new rocks.

Slate is a dull grey rock formed when sedimentary mudstone was compressed deep underground. Mudstone grains grew into flaky crystals which were arranged in parallel layers. This is why slate splits easily into flat sheets. These have been used for many years in the building industry as roofing material though today they are expensive to produce and have mostly been replaced by manufactured clay or cement tiles.

Questions

1 What is metamorphic rock?

2 Explain how marble is made.

3 Why are some statues made from marble and not limestone?

4 Explain why very old marble statues, like those in Greece, are not quite like they were when they were made.
(*Hint*: remember that rain water is slightly acid.)

5 How was slate made?

Marble statues, badly affected by air pollution

And so the rock cycle is complete

When you look at a mountain or a large cliff face it is easy to think that rocks are permanent and will never change. However, it is important to realize that rocks play a temporary part in the evolution of our planet. Igneous rocks are weathered and eroded. The debris is transported then dumped to eventually become sedimentary rock. This may be buried deeply and changed by heat and pressure (metamorphism) and then be lifted to form a mountain range only to be eroded again! The rock cycle goes on and on . . .

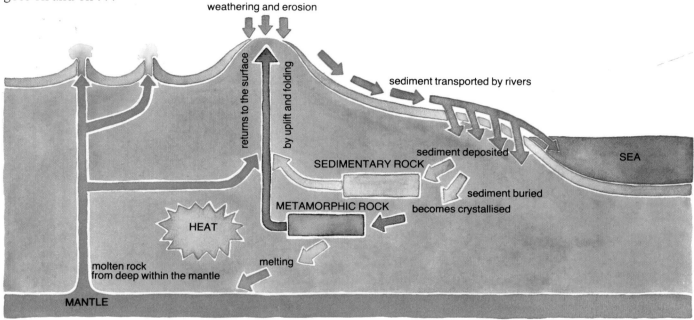

The rock cycle

'The present is the key to the past'

In the late eighteenth century James Hutton put forward one of the most important ideas in earth science. It was called **uniformitarianism**.

Hutton believed that we can understand the geological history of the Earth surface by studying the processes that we can see happening today – the rock cycle!

Questions

1 Explain why the rock cycle is called a **cycle**.

2 In theory, material from the mantle could reach the surface and then be reburied until it became molten again.
    a) Describe briefly how this could happen.
    b) Roughly how long do you think the cycle would take?

3 Why is water important in the rock cycle?

4 'All mountains we see today were formed at the beginning of the Earth and will never change.'
Do you agree with this statement? Explain your answer.

Rocks for building

If you look carefully at the older buildings in your town or village you will probably notice that they are built of stone. For many years builders used local stone, dug from nearby quarries, because at the time it was readily available and cheap.

Rock that could be cut or split in any direction was the most favoured by builders. The sedimentary rocks sandstone and limestone are good examples. They are soft and so stonemasons could '**dress**' or shape stones to make sharp corners, curved arches or even ornate carvings. In some parts of the country granite was used. This is a hard igneous rock and very difficult to shape. However, granite is very hard wearing and not attacked by chemicals in the atmosphere. This explains why so many churches and public buildings still look so good after standing for hundreds of years.

For roofing, the metamorphic rock, slate, was often used. Most of the slate came from quarries in Wales or the Lake District in England. Slate is easily split into thin sheets and these were cut for fixing in an overlapping pattern onto roofs.

Flint is a very hard stone which can be chipped but not cut. In some areas, houses and churches were built using flints stuck together with builder's mortar. Because the flints could not be 'dressed' the corners of the building were built in brick or stone.

Granite is an expensive but long-lasting building material.

Slate roofs; clay tiles are now used instead

Delicate carving needs soft stone

Traditional flint and brick building

Clay for building

Stone buildings last well but stone is difficult to quarry, hard to cut and expensive to transport. The rock that plays the biggest part in the modern building industry is **clay**.

Clay is soft and pliable when wet, as you may have noticed when making pots or models. When it is heated strongly it becomes hard and strong. This makes it an ideal material for bricks and roof tiles.

Cement

Clay is also an essential part of cement. Ground-up clay and limestone are roasted together in kilns then ground up into a fine powder. Builders use cement mixed with sand and water to make mortar to stick bricks together. Concrete, another popular building material, is a similar mixture to mortar but this time gravel is added to increase strength.

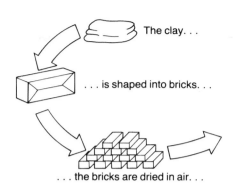

The clay. . .

. . . is shaped into bricks. . .

. . . the bricks are dried in air. . .

Making bricks

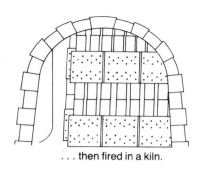

. . . then fired in a kiln.

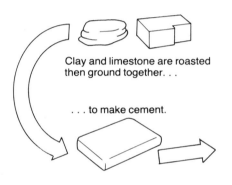

Clay and limestone are roasted then ground together. . .

. . . to make cement.

Making cement

Mixing the cement with sand, gravel and water makes concrete.

Activities

Make a survey of some buildings in your town. List the types of rock used in their construction. Try to answer these questions for each building.

What is the building used for?

What are the walls made of?

Are the corners of the building or the doorways made of different stone?

Are there any stone or clay decorative features (carvings, arches, etc.)?

What is the roofing material (clay tiles, slates, other)?

Choose five or more buildings for your survey. Include all of the following: your home, your school, an old church, a public building such as the town hall or library, an old house.

This activity can be extended by keeping notes on house building when you travel on holiday or by taking photographs of interesting buildings.

Questions

1 Give one advantage and one disadvantage of the following building materials: **a)** granite **b)** limestone.

2 Why do you think that builders used local stones?

3 Why are new houses not usually built of stone? Describe how bricks are made.

4 Explain why slate was used as a roofing material all over the country.
Why do you think that clay or concrete tiles are now used?

5 In what way is the firing of bricks like metamorphism?

6 The road shown in the photograph is covered in granite chippings. Can you think why?

Tectonic processes: what are they?

Volcanic rocks and sediments were originally laid down in horizontal layers or **strata**. However, if you look at rocks in quarries or cliff faces you may notice that the rocks are tilted, bent or fractured. They are no longer horizontal. These rocks have been deformed by mechanical forces or **tectonic processes**. (Tectonic comes from the Greek word for carpenter – someone who shapes wood and makes new structures).

Four things influence the way in which rock is deformed:

- the kind of rock it is, some are more plastic than others.
- the temperature – rock is more plastic at high temperatures.
- the force applied, whether it is a compression (push) or tension (pull).
- the length of time that the forces act.

You can model these effects by using pieces of Plasticine.

The greater the time that a force is applied the greater the change in shape of the rock layers. If rock is kept under a constant force it will continue to move or **creep**. Creep causes problems for engineers building large structures such as bridges.

A suspension bridge

At high temperatures, tension causes stretching.

At high temperatures, compression causes rocks to bulge.

Questions

1 **a)** Explain why new layers of rock are usually horizontal.
   b) Why are the layers of rock in cliff faces often tilted or broken?

2 What is **a)** compression **b)** tension?

3 Describe, using diagrams, the effects that
   a) compression **b)** tension forces have on rock deformation.

4 List some examples of structures (e.g. bridges) that may be affected by 'creep'.

Joints and faults

Fractures, or breaks, in layers of rocks are often due to tension. They are usually found in igneous rocks. Hot volcanic rocks shrink as they cool. This sets up pulling forces in the rock causing it to break up into a clear pattern of fractures or **joints**.

Joints are very common in all kinds of rocks and are simply cracks in a piece of rock where it has split but where the pieces have not moved apart.

The Giant's Causeway in Antrim, Northern Ireland is a marvellous example of jointing. The basalt (an igneous rock) has cooled, shrunk and produced vertical joints. These outline perfect hexagonal columns.

The granite tors of Dartmoor are also the result of jointing, this time rectangular blocks are produced.

Faults are formed when movement takes place along a fracture line in a piece of rock. Because the rocks on one side of the fracture line move, they no longer line up with the rocks on the other side.

There are various kinds of faults. The most common is a **normal fault** like the one shown in the photograph below.

Giant's Causeway, County Antrim

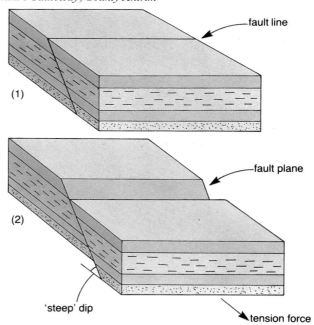

A normal fault. Movement takes place along a fault plane. The angle between the fault plane and the horizontal is called the dip.

A normal fault. The rocks on the right have moved down in relation to those on the left. You should be able to match up the layers on either side.

Questions

1 What is the difference between a *joint* and a *fault*?

2 Why are joints usually associated with igneous rocks?

3 Under what conditions would you expect joints to form in sedimentary rocks? (You may need to look back to page 63

4 Suggest how the tors on Dartmoor were formed.

5 Describe how the fault shown in the diagram above could have been formed.

Folds in the Earth's rocks

If a rock is compressed while it is still soft and plastic it will **fold** instead of cracking. There are two kinds of folds:

- upfolds are called **anticlines**.
- downfolds are called **synclines**.

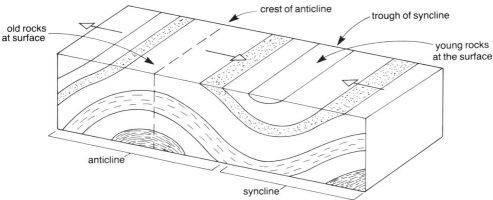

Folded rocks!

Notice how the surface rocks are younger near the middle of a syncline and older near the middle of an anticline.

Rocks usually crack open as they are folded upwards, just like when you bend a rubber, and this exposes the older rock. Joints produced in this way are often eroded by weathering.

It is possible to see folding (on a much smaller scale) nearer to home. This photograph shows folding in the sedimentary rocks at Stair Hole, Lulworth Cove, Dorset.

The highest mountains on Earth have been formed as a result of folding. The Alps, Andes, Himalayas and Rockies, are all folded mountains. The photograph shows Mount Everest.

Questions

1 What condition must rocks be in to fold?

2 What is **a)** an anticline **b)** a syncline?

3 Why do we usually find older rocks exposed at the surface above an anticline?

Tear faults and earthquakes

Faults and folding result in vertical movements of rock. Other types of fault, called tear faults, allow rocks to move horizontally on either side of a fracture.

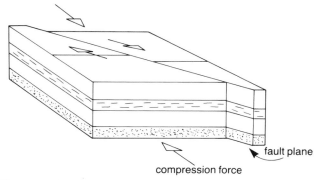

A sliding or tear fault

Unfortunately these movements are rarely smooth. As compression forces build up, a time is reached when something has to give. The rocks suddenly jerk past one another releasing energy as heat, sound and, more importantly, **shock** or **seismic waves**. We feel these as an earthquake.

Seismic waves are produced at a point called the focus, deep within the Earth. The point on the Earth's surface above the focus is called the **epicentre**.

Most major earthquakes appear in well defined seismic zones called 'earthquake belts'. The seismic zones mark the edges of huge mobile pieces of the Earth's crust. These pieces are called **plates**.

The best known tear fault is the San Andreas fault in California. The last major earthquake in 1906 destroyed much of the city of San Francisco. Rocks slid 7 metres past each other over a distance of 300 kilometres. Compression forces have been accumulating ever since and experts are predicting another major earthquake soon!

The seismic waves produced during earthquakes provide valuable evidence about the structure of the inside of the Earth. The speed at which seismic waves travel depends upon the type of the material they are passing through. By measuring the speeds of seismic waves, scientists have produced a model of the Earth's interior, like that shown on page 59.

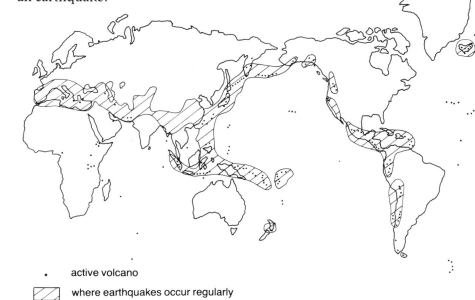

- active volcano
- where earthquakes occur regularly

The world's earthquake belts

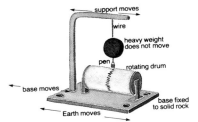

A seismometer measures and records the size of movements in the Earth's crust due to earthquakes and tremors.

Questions

1 What causes earthquakes?

2 What are seismic waves? How have they helped earth scientists produce a model for the structure of Earth?

3 Does Britain lie in a seismic zone? Name one country that does.

4 Suggest the function of the heavy weight in the seismometer shown in the diagram?

5 Even the smallest shock waves near the San Andreas fault are recorded. Why are these important for **a)** geologists **b)** people who live near the San Andreas fault?

Plate tectonics

The theory of plate tectonics suggests that the Earth's crust consists of large rigid plates of rock floating on the molten mantle. These plates are built up at ocean ridges and destroyed as they are forced beneath continents. It is believed that convection currents in the mantle cause the plates to move.

The Mid Atlantic Ridge – a constructive plate margin

Beneath the Atlantic Ocean, molten magma rises from the mantle below. This magma forms a ridge and creates new ocean floor. This in turn pushes the plates on either side away from each other and causes them to spread outwards. The rate of spreading can be calculated and is estimated to be up to 8 cm per year.

The west coast of South America – a destructive plate margin

For over a hundred million years South America has been moving slowly away from Africa and colliding with the floor of the Pacific Ocean. As the two plates meet the ocean floor is forced down into the mantle. The ocean floor gets hotter and melts to form magma. Magma rises up through the mantle and melts the continental crust above. Eventually magma may erupt through the surface to form a volcano.

1 *Magma rises from the mantle by convection.*
2 *New ocean floor spreads outwards.*
3 *The floor of the Pacific Ocean is forced beneath South America. The ocean crust melts to form magma.*
4 *Volcano formed by rising magma.*

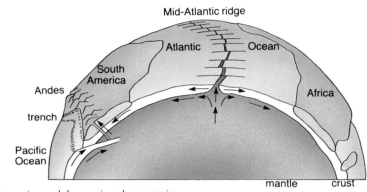
Constructive and destructive plate margins

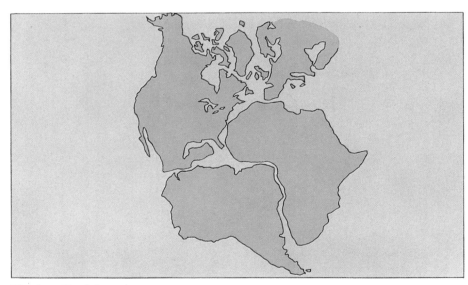
The planet Earth jigsaw!

Scientific evidence suggests therefore that the rocks of the ocean floor and the continents move slowly over the surface of the Earth. So it seems likely that the shape and positions of the continents have changed through time. There is a lot of support for this idea of **continental drift**. Earth scientists have been able to match up the rock types on either side of the Atlantic Ocean, some sedimentary rocks in South America and Africa even contain the same fossils. Even the present day shapes of the continents match, rather like a huge jig-saw puzzle.

Questions

1 Explain as simply as you can the theory of plate tectonics.

2 What happens at **a)** a constructive plate margin **b)** a destructive plate margin?

3 What is continental drift?

4 What evidence is there for the idea of continental drift?

5 Draw a diagram like the one above to show what might happen when two continents eventually collide.

1 Two students are talking.

Student *A*: The Earth is a solid ball of rock.

Student *B*: The Earth has a solid crust but the inside is full of molten rock.

a) What evidence would you use to show Student *A* that the Earth is not solid?

b) Draw a labelled diagram to show Student *B* what geologists believe the Earth is like inside.

2 The table below shows some of the rocks found in the United Kingdom.

place	rock	period
Edinburgh	basalt	Carbon-iferous
North Wales	slate	Cambrian
Yorkshire	coal	Carboniferous
Cotswolds	limestone	Jurassic

a) Granite is an **igneous** rock. Explain how it was formed.

b) Roughly how old are the mountains of North Wales?

c) Dinosaurs lived between 225 million and 65 millions years ago.

 i) Why don't we find dinosaur fossils in coal?

ii) Where might you find dinosaur fossils in England?

3 The diagram shows a house.

flint — slates — brick

a) Suggest **two** reasons why the builders chose flint for the walls.

b) Suggest why the builders chose bricks for the corners of the building.

c) Suggest one reason why the builders chose slate for the roof.

4 The diagrams below show how our continents look now and how they might have looked 200 million years ago.

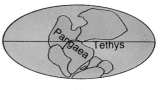
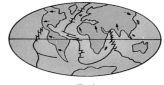

200 My ago   Today

a) Explain using the theory of plate tectonics how our continents have moved so far apart.

b) What evidence do scientists have for the theory of plate tectonics?

c) The distance between Africa and South America is roughly 7000 km. Estimate the average speed at which the continents are moving apart.

5 The graph shows the range of size of rock particles that can be   i) picked up and   ii) deposited by different water current speeds.

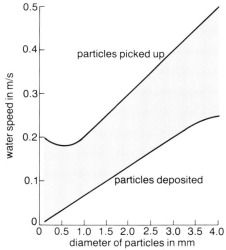

a) What water speed is needed to pick up particles of 3 mm diameter?

b) At what water speed will particles of 1.5 mm diameter be deposited?

c) What does the shaded area represent?

d) What is the relationship between the speed of water flow and the size of rock particles carried?

e) What range of particle sizes would you expect to be deposited by a river whose water speed suddenly changed from 0.2 m/s to 0.02 m/s?

4 Fuels and energy

What is a 'fuel'?
Why does burning fossil fuels cause problems?
How can we conserve fuel supplies?
What alternative energy sources can we use?

Since prehistoric times humans have burnt things to give them heat. Burning wood kept our ancestors warm at night and let them cook their food. The energy from the fires was also used to 'fire' clay pots and make them hard. Metals for weapons and for jewellery were produced in their wood and charcoal fires.

Modern civilizations depend on fuels to provide energy for transport, industry, and for use in the home. In some parts of the world wood is still a very important fuel but better fuels have been discovered and developed. Scientists can also 'design' fuels for special purposes.

The pictures below show some uses of common fuels.

Coal fire. In some parts of the United Kingdom smokeless fuel must be used in place of coal.

This cooker burns natural gas which is piped to the house.

This train uses diesel oil as a fuel. Many ships, lorries, and other large vehicles have diesel engines.

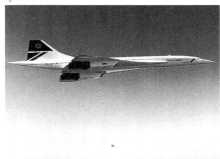
Like diesel oil and petrol, aviation fuel is produced from crude oil.

Petrol is burnt in the engine of this car. It is a liquid fuel produced from crude oil.

Bottled gas. Propane and butane can be stored under pressure in metal cylinders as liquids. This gives a portable fuel supply.

Questions

1 Name two solid fuels, two liquid fuels, and two gaseous fuels.

2 What fuels, if any, are burnt in your home **a)** to keep it warm **b)** to cook food? (If your answer is 'none', explain why.)

3 If someone in your family has a car, what fuel does it use?

4 Describe two situations where 'bottled gas' is much more convenient than natural gas.

5 Some garages sell a fuel called 'DERV'. What is 'DERV'? What type of vehicles use this fuel?

6 Some people believe that if humans had not learnt how to use fire we would have become extinct millions of years ago. How do you think fire helped our ancestors to survive?

Making better fuels

In prehistoric times humans used wood as a fuel. It was plentiful, easy to carry, and it burnt well. It did not really matter that the fires were smoky or that they left a lot of ash. But it was soon discovered that a much better fuel can be made from the wood. It is called **charcoal**.

Wood fires are smoky because the wood contains impurities. Most of these can be driven off as liquids and gases by heating the wood carefully. Of course, if the wood is heated in air it catches fire. To make charcoal it must be heated in a closed container so that very little air can get in.

This process is called **distillation**. The distillation of wood is said to be the oldest chemical process used by humans.

Distillation of wood in the laboratory

Thin strips of wood can be turned into charcoal by heating them as shown in the diagram below. As the wood is heated gases are given off. Some of these can burn. Others can be cooled and turned into an oily mixture of liquids. This is a useful by-product. A form of alcohol can be extracted from it, and also creosote, which can be used to preserve wood (fences, for example).

The black substance left in the boiling tube is charcoal. We can think of it as being nearly pure carbon.

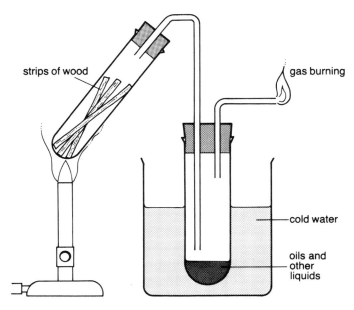

Distillation of wood in the laboratory

Charcoal as a fuel

Charcoal is a much better fuel than wood. It gives more heat for the same weight of fuel. It burns cleanly with very little smoke. It also gives very high temperatures particularly if air is blown through it while it is burning.

Charcoal is used for barbeques. It burns cleanly and gives a steady, high temperature for cooking. A flammable liquid can be used to start the charcoal burning. This gives enough energy to start the reaction then the heat given out by the burning charcoal keeps it going.

Problems of charcoal burning

The use of charcoal as a fuel causes problems in some developing countries. Many villagers cut down trees to make charcoal. Cutting down the trees means that there is less shade from the sun and so moisture evaporates from the soil. The roots of the tree no longer hold the top soil together and it blows away. Soon the area becomes desert and the villagers have to move away. In many countries, trees are being planted to stop the spread of the desert but the people will cut down the new trees for wood and for charcoal unless they are given alternative fuel supplies.

Formation of coal

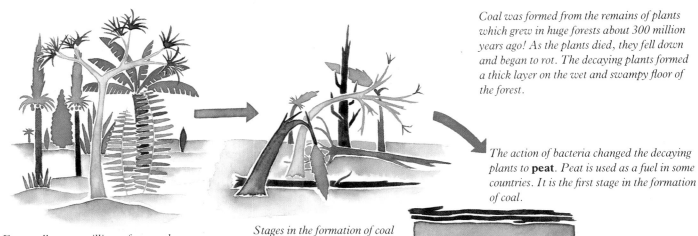

Coal was formed from the remains of plants which grew in huge forests about 300 million years ago! As the plants died, they fell down and began to rot. The decaying plants formed a thick layer on the wet and swampy floor of the forest.

The action of bacteria changed the decaying plants to **peat**. Peat is used as a fuel in some countries. It is the first stage in the formation of coal.

Stages in the formation of coal

Eventually, over millions of years, the decaying plants were changed into coal.

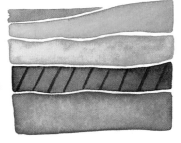

Gradually the land sank and water covered it. Layers of mud and gravel were laid over the decaying plants. As more and more rocks were laid down by the seas above, the pressure on the peat layers increased. The temperature also got higher.

What is coal?

We can think of coal as an impure form of the element carbon. There are several different types of coal. These were formed at different pressures and temperatures under the Earth's surface.

type of coal	approximate amount of carbon	description	main mining areas in the UK
anthracite	90%	Hard, black coal. Formed at great depth.	South Wales
bituminous	60%	Formed at lower pressure than anthracite.	Nottingham and Yorkshire
lignite	40%	Softer coals, sometimes brown in colour.	

Fossil fuels

Coal is called a **fossil fuel** because it was formed from living things. Fossils of plants can sometimes be found in lumps of coal.

Natural gas and oil are also fossil fuels (see pages 80 and 82). The uranium used as a fuel in nuclear power stations is not a fossil fuel. It is a naturally occurring element.

Questions

1 How do we know that ferns were alive over 300 million years ago?

2 Explain why human fossils are not found in coal.

3 How were the layers of plant material buried under mud and gravel?

4 How could you tell the difference between a piece of anthracite and a piece of lignite?

5 Why is anthracite a better fuel than lignite?

6 Suggest why the peat found in Ireland has not turned to coal.

7 Why do you think that geologists call the time when coal was being formed the **carboniferous period**?

Burning carbon

Wood, charcoal and coal are all fuels which contain **carbon**. Carbon is a chemical element found in all living things. It is also found in minerals like coal. There are several different forms of carbon. The **graphite** inside a pencil is a form of carbon and so is **diamond**. The diagrams opposite show how we can investigate the **combustion** or burning of carbon.

When powdered carbon is heated it glows red hot as it reacts with oxygen in the air. When burning carbon is put into pure oxygen gas it flares up with sparks and a bright flame. The bright flame and hot sparks show us that the burning carbon is giving out heat. After the flame goes out, the gas produced by burning carbon can be tested.

Some calcium hydroxide solution (also called lime water) is poured into the gas jar. The clear liquid turns a milky white colour. This test tells us that the gas is carbon dioxide.

The equation for the combustion of carbon is:

carbon + oxygen → carbon dioxide + heat

Using symbols:

$$C + O_2 \rightarrow CO_2 + heat$$

It is this reaction which makes charcoal and coal good fuels. It is an **exothermic** reaction. Exothermic means that heat is given out.

The gas in the gas jar can also be tested by putting some wet, blue litmus paper in the top of the jar. It turns a purple-red colour showing that carbon dioxide gas dissolves in water to give a weak acid.

carbon dioxide + water → carbonic acid

Using symbols:

$$CO_2 + H_2O \rightarrow H_2CO_3$$

Incomplete combustion

Usually when carbon burns carbon dioxide is formed. If, for some reason, there is not enough oxygen available, another gas is produced. This is **carbon monoxide**. The diagrams opposite represent molecules of the two gases. The equation below is for carbon burning in a limited supply of oxygen. Notice how two carbon atoms react with just one oxygen molecule.

carbon + oxygen → carbon monoxide
$$2C + O_2 \rightarrow 2CO$$

Remember to wear safety glasses!

long-handled combustion spoon

lid

oxygen gas

Burning carbon in oxygen

colourless calcium hydroxide solution ('lime water') goes... . milky white

moist, blue litmus goes... purple/red

carbon dioxide gas

Carbon dioxide gas is acidic. It gives a white precipitate with calcium hydroxide solution.

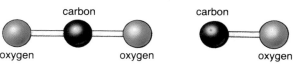

carbon

oxygen   oxygen

carbon

oxygen

One molecule of carbon dioxide (CO_2)   *One molecule of carbon monoxide (CO)*

Car engines give out large amounts of carbon monoxide gas. This is because in the engine the fuel (petrol) is being burnt in a very limited amount of air. This is not usually a problem because once the carbon monoxide leaves the exhaust pipe it is quickly 'diluted' in the air. However, in a confined space carbon monoxide is very dangerous because it is a highly poisonous gas.

It is extremely dangerous to run the engine of a car or motorbike inside a closed garage. The carbon monoxide gas is colourless and does not smell. It can quickly make a person unconscious. If they continue to breathe the gas they will die.

Questions

1 Describe these forms of carbon:
  a) coal **b)** diamond **c)** graphite.

2 What colour does wet, blue litmus paper turn in carbon dioxide gas? Why does the litmus paper have to be wet?

3 Why do long road tunnels have large ventilation fans? What might happen if they didn't?

Ash, smoke, and soot

When pure carbon burns completely in oxygen the only product is carbon dioxide gas. There are no solid products left behind. Coal is not pure carbon. It contains some impurities which do not burn and others which burn to give solid products. These are left as **ash** when coal burns.

Large pieces of ash block up the fireplace or boiler where the coal is being burnt. This stops air getting to the unburnt fuel and so it does not burn as well. The ash has to be regularly cleared away.

Good fuels do not leave much ash.

Very small particles of ash are carried out of the chimney as **smoke**. Smoke and ash pollution causes damage to buildings. The smoke and soot blackens the buildings. Buildings can be cleaned by blasting the stone with sand, but this is very expensive.

About 50 years ago in Britain, nearly all factories and power stations burnt coal. Steam trains used coal fires to heat their water and people burnt coal to keep their homes warm. All these things made smoke. In industrial areas and in large towns this caused serious pollution. London and other major cities suffered from **smog**.

Smog is a word made up from **sm**oke and **f**og. Smog was like dirty fog. Tiny pieces of ash and soot hung suspended in the damp air making it difficult to see far. The smog also damaged the people's health because the smoke particles irritated the lungs. Old people, young children and people suffering from asthma or bronchitis were at risk. In fact more than 3000 people died in London in 1952 due to heavy smog.

Eventually the problems caused by burning coal got so bad that in 1956 a law was passed called the Clean Air Act. This made it illegal to burn coal in some areas. These areas are called **smokeless zones**. In them people have to use electricity or gas or they have to use 'smokeless fuels'. These are fuels made from coal but which do not make much ash or smoke.

Since the Act was passed, the air in our cities has become much cleaner. We do not have smogs caused by coal burning anymore. Some large cities in other countries now suffer from smog produced by car exhaust fumes!

London smog was dirty and unpleasant. It also caused hundreds of deaths.

Activities

1 Find out whether you live in a smokeless zone or not.

2 Try to get samples of 'real' coal and smokeless fuels. (Your local coal merchant may be willing to help.) Compare the fuels. What do they look like? What do they feel like? How much do they cost?

3 If you live in a town, make a list of any buildings which have been blackened by smoke pollution. Make a list of any buildings which have been cleaned by sand blasting.

Questions

1 What is ash?

2 Why doesn't pure carbon leave ash behind when it burns?

3 Soot is a fine black powder. It will burn to give a gas which turns calcium hydroxide solution milky white. What element is soot? Explain how you know.

4 What is 'smog'?

5 In heavy smog, people used to wear cotton masks over their mouths and noses. Why was this?

6 Explain why London has not had a serious smog since about 1956.

Making a better fuel from coal

We have already seen how charcoal can be made from wood by distillation. The same method can be used to make a better fuel from coal.

The coal is heated in a closed container. Because air cannot get in the coal does not burn. Instead the heat drives off the impurities as gases. A very important fuel called **coke** is left behind.

The gases and liquids collected as by-products during the distillation are very useful.

Distillation of coal in the laboratory

The diagram shows how coke can be made in the laboratory. The coal should be heated fairly gently at first and then more strongly. After several minutes the boiling tube can be allowed to cool. The contents should then be scraped out onto a fire-proof mat and compared with some coal. The coke is black but not so shiny as the coal. It also has a rougher texture.

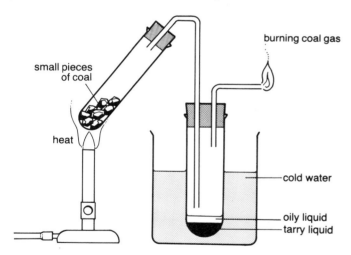

Distillation of coal

While the coal is being heated, gases are given off. Some of the gases will burn at a jet and so can be used as a fuel. In fact coal gas was used to heat homes and for street lighting for many years. It has now been replaced by electricity and natural gas.

The other gases given off by the coal cool down and turn into liquids at the bottom of the cold test tube.

The lighter liquid on the top can be tested with red litmus paper. This turns blue showing that the liquid is alkaline. Ammonia, a very important chemical, can be extracted from the liquid.

Products from coal

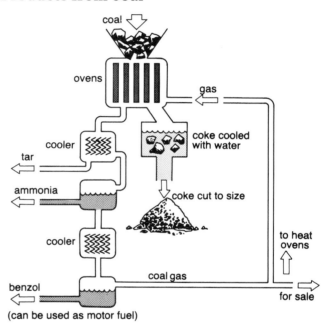

A modern coking plant. Some of the coke will be used as a smokeless fuel. The rest will be used to make steel.

Questions

1 Name two fuels produced by distillation.

2 Why isn't air allowed into the oven when making coke?

3 When 10 kg of coal are heated, 7 kg of coke are left. Where have the other 3 kg gone?

4 Coal makes lots of smoke when it is burning. Coke is a smokeless fuel. Explain why.

5 State two uses of coke.

6 What is a 'by-product'? Name four by-products of making coke from coal.

Other fossil fuels

Hundreds of millions of years ago, while ancient forests were starting to make coal on the land, other fossil fuels were being made in the sea. We use them now as **oil** and **natural gas**.

Oil was made from the microscopic plants and animals which lived in the sea. As they died their bodies collected at the bottom of the ocean. Here they were covered by mud and sand. Over thousands of years the layers of mud and sand became very thick. This put the decaying plants and animals under pressure. High temperatures and the pressure changed them into a thick black liquid called crude oil.

The layers of sand and mud turned into rocks. We call rocks laid down by the sea, **sedimentary rocks** (see page 63). These have tiny holes in them rather like a sponge. The holes allow liquids to slowly move through the rock. The rock is **porous**.

The crude oil moved slowly through the porous rock. However, in some places the layers of rock had become folded so that the oil was trapped under solid, hard rock. The oil could not move through this **impermeable rock** and so it collected together. It is these 'pockets' of trapped oil that oil companies look for.

How do we find oil?

Large pockets of crude oil are found where layers of porous rock meet impermeable rock. We are more likely to find oil where the layers have been folded to make a 'dome' shape. Oil geologists study geological maps to find areas where the right types of rocks are found. They can then drill deep holes and remove samples of rock. This tells them what the rocks under the ground are like.

To check the shape of the rock layers, the geologists can send shock waves through the ground. They do this by setting off an explosion. They then use special microphones to record any echoes. By studying the echoes received at each microphone, the shape of the rock layers can be mapped.

All these methods are used to find areas where the oil company is **likely** to find oil or gas. The only way to find out for certain is to drill a test hole. Sometimes several bore holes will be needed before oil is found. Sometimes no oil or gas is found. Exploring for oil is very expensive!

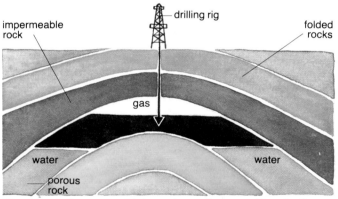

The gas and oil are trapped below a dome of impermeable rock. Once the drill gets through this rock, gas and oil will be released.

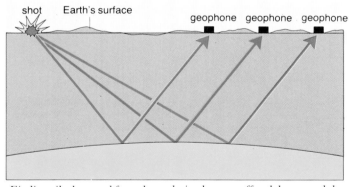

Finding oil: the sound from the explosion bounces off rock layers and the echoes are picked up by special microphones called geophones.

Questions

1 Explain why crude oil is called a 'fossil' fuel?

2 What does crude oil look like?

3 What does 'porous' mean?

4 Design an experiment to show that a stick of chalk is porous.

5 What does 'impermeable' mean?

6 What does 'geology' mean?

Fuels and other things from oil

When crude oil is pumped from the ground it is a black, smelly liquid. It is a mixture of many chemicals which, on their own, are very useful. They can be separated by a technique called **fractional distillation**. The crude oil is heated to a temperature of about 400°C. This turns it into a mixture of gases which are pumped into a tall tower. The hot gases rise up the tower.

The top is cooler than the bottom, so as the gases move upwards some of them start to turn back into liquids. Because they have different boiling points the different liquids form, and are pumped out at different heights. The diagram shows how the crude oil is separated into fractions in the fractionating tower. Notice how gases are taken off at the top, thin liquids in the middle, and thick liquids and waxes at the bottom.

Oil fractionating plant

Using the fractions

Crude oil has become one of the most important sources of **organic compounds**. Organic compounds are compounds which contain carbon atoms joined to hydrogen and sometimes other atoms. They can be very simple like methane or can contain long chains or rings of carbon atoms. These compounds are very important to industry and in our everyday lives. Crude oil supplies hundreds of organic compounds. These are used in man-made textiles like nylon and Terylene, in explosives, in paints and dyes, in cosmetics, in plastics, in car tyres, in road surfaces . . . the list is almost endless. In addition to these things, fractional distillation of crude oil gives us a number of different fuels.

Petroleum gas

Petroleum gas is a mixture of light organic compounds called alkanes (see page 82). These can be turned into liquids and stored inside metal cylinders. This is useful because it gives us a portable fuel supply. Some camping stoves used bottled gas. Propane gas can be used to run gas central heating in areas where there is no supply of natural gas.

Petrol and paraffin

Petrol and paraffin are liquid fuels. Petrol is used for car engines. It contains molecules with between 5 and 10 carbon atoms. Octane is an organic compound with 8 carbon atoms. Look for this name when you are next in a garage. Paraffin is no longer a popular fuel for use in the home but some types of greenhouse heaters burn it. A special type of paraffin called kerosene can be used in the engines of jet aircraft.

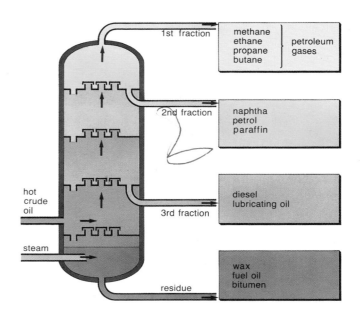

Diesel oil

Diesel oil is used for vehicles with a special type of engine. Diesel engines are very reliable but are noisier than petrol engines. Tractors, lorries, buses and ships usually have diesel engines.

Fuel oil

Fuel oil is a thick liquid which can be burnt in the boilers used to heat schools, factories and other large buildings. Some power stations use fuel oil to produce electricity.

Natural gas is a good fuel. It can be piped to people's homes for cooking and heating. It burns cleanly and does not leave any ash.

What is natural gas?

Natural gas is a compound called methane. Methane is a hydrocarbon – a compound made up of the elements carbon and hydrogen. Carbon atoms can make four **bonds**. In methane the carbon atom is bonded to four hydrogen atoms as shown in the diagram. Methane has the formula CH_4.

The combustion of methane can be investigated using the apparatus shown opposite.

The funnel above the burner collects any products from the combustion and the pump pulls these through the apparatus.

The anhydrous copper sulphate powder in the first test tube is white when the investigation starts but eventually turns blue. This shows us that water is present.

The calcium hydroxide solution in the second test tube starts as a clear liquid but turns cloudy and white as the gases from the burning gas bubble through. This tells us that carbon dioxide is produced (see page 77).

The reaction for the burning of methane can be written:

methane + oxygen → carbon dioxide + water + heat

Using symbols:

CH_4 (g) + $2O_2$ (g) → CO_2 (g) + $2H_2O$ (g) + heat

The symbol (g) shows that the substance is a gas. We can see therefore that when methane burns in lots of air the only products are gases – carbon dioxide and water vapour. There is no ash and no 'pollution'.

Other fuels like methane

Methane is one of a 'family' of carbon compounds which can be used as fuels. The structures of some of them are shown opposite.

Notice that each carbon atom is joined to four other atoms. Each bond is either between two carbon atoms or between a carbon and a hydrogen. Compounds with this type of structure are called **alkanes**. Most alkanes can be used as fuels. Many of them can be extracted from crude oil.

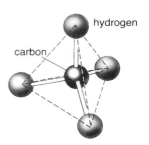

A molecule of methane. The hydrogen atoms make a triangular pyramid. The carbon atom is at the centre of the tetrahedron (pyramid).

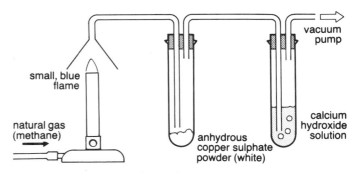

Investigating the burning of natural gas

CH_4	C_2H_6	C_3H_8	C_4H_{10}
methane	ethane	propane	butane

Alkanes

Questions

1 Octane is an alkane with eight carbon atoms in each molecule. Draw the structure for octane.

2 In which common fuel would you find octane?

3 The first four alkanes have the formulae CH_4, C_2H_6, C_3H_8, C_4H_{10}.

a) Write down in words or in mathematical symbols the pattern between the number of carbon atoms and the number of hydrogen atoms in an alkane.

b) Use your answer to work out the formula for the alkane with 14 carbon atoms.

Burning fossil fuels ... more problems

Fossil fuels like coal and oil contain the element **sulphur** as an impurity. When they burn, the sulphur reacts with the oxygen in the air to make sulphur dioxide. This gas can cause serious pollution in the atmosphere.

Burning sulphur

Sulphur is a bright yellow element. When it is heated in air it starts to burn with a blue flame. The gas given off has a strong, choking smell and so this investigation is best done in a fume cabinet.

When burning sulphur is put into pure oxygen gas it burns with a bright blue flame. The gas given off is called sulphur dioxide.

$$\text{sulphur} + \text{oxygen} \rightarrow \text{sulphur dioxide}$$
$$S\,(s) \quad + \; O_2\,(g) \rightarrow \qquad SO_2\,(g)$$

The sulphur dioxide trapped in the gas jar can be dissolved in a little distilled water and then tested with indicators. When a piece of blue litmus paper is dipped in the solution it turns red. When Universal indicator solution is added to the solution it turns red. **These results show that sulphur dioxide gas dissolves in water to give a strong acid**.

The acid produced when sulphur dioxide is added to water is called sulphurous acid.

$$\text{sulphur dioxide} + \; \text{water} \; \rightarrow \text{sulphurous acid}$$
$$SO_2\,(g) \qquad + H_2O\,(l) \rightarrow \quad H_2SO_3\,(aq)$$

Sulphur dioxide pollution

Industries and power stations that burn coal, coke, or oil let sulphur dioxide escape from their chimneys. This mixes with the air and moves around as the wind blows. The sulphur dioxide dissolves in water in the air and so when it rains the rain is acidic.

Sulphur dioxide is not the only acidic gas in the atmosphere. Petrol engines give out nitrogen dioxide in their exhaust gases. This also dissolves in water to give acidic rain water. **Acid rain is a serious problem in many parts of the world**.

The effects of acid rain

The effects of acid rain on plants has a serious effect on the environment. If trees die because of acid rain then the woodland animals may also die. In fact the balance may be so upset that the environment may be changed forever.

Acid rain also causes problems in lakes. The rain makes the water in the lake slightly acidic. Some of the small organisms in the lake cannot live in acidic conditions so they die. Larger organisms, for example fish, then have less food and so some of them die. Their decaying bodies cause more pollution until the lake has very little life in it.

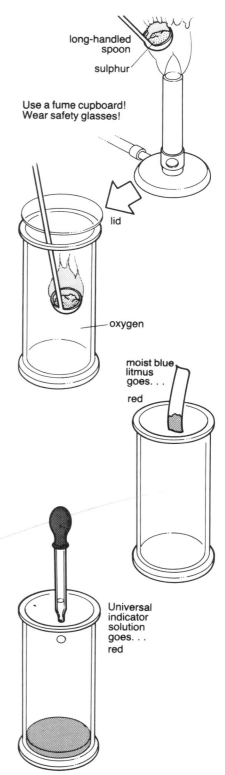

long-handled spoon

sulphur

Use a fume cupboard! Wear safety glasses!

lid

oxygen

moist blue litmus goes... red

Universal indicator solution goes... red

Burning sulphur in oxygen: sulphur dioxide gas is acidic. It has a strong, choking smell. This experiment should be done in a fume cupboard.

Comparing fuels

We can compare fuels by measuring how much energy is released when 1 g of the fuel burns. The diagram shows how we can do this for butane gas.

The burner with the fuel inside is weighed before the investigation starts. The starting temperature of the water is also measured.

The flame is then lit. As the water gets hotter it is stirred so that it is all at the same temperature. When the water is about 40 °C hotter than at the start, the flame is put out. The accurate final temperature is then measured.

Finally the burner with the fuel left inside is weighed. This can be used to find the mass of the fuel used to heat the water. The table shows how the results can be recorded.

The investigation can then be repeated with other fuels.

How much energy?

The results from the investigation help us to compare fuels but some more calculation is needed if we want to find out how much energy the fuel has released.

We need to know two things:
4200 joules of energy heats up 1000 g of water by 1°C.
1 cm^3 of water has a mass of 1 g.
We can now do the calculation:
To heat up 1000 g of water by 1°C needs 4200 J of energy.
The investigation used just 500 g of water so . . .
To heat up 500 g of water by 1°C needs 2100 J of energy.
The water in the investigation was heated up by 40 °C so . . .
to heat up 500 g of water by 40 °C needs 2100 × 40 J of energy.
84 000 J or 84 kJ heats up 500 g of water by 40 °C.

We burnt 3 g of butane so . . .
3 g of butane releases 84 kJ of energy.
1 g of butane would release 84/3 kJ = 28 kJ.

This is a useful figure for making comparisons but is not very accurate. This is because some of the fuel's energy went to heating the beaker and the air around the apparatus. Accurate measurements show that 1 g of butane releases about 50 kJ of energy when it burns.

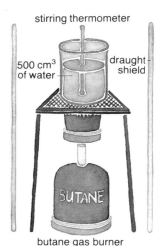

Energy from burning butane

Name of fuel: butane gas	
Mass of burner + fuel *before*	327 g
Mass of burner + fuel *after*	324 g
Mass of fuel *used*	3 g
temperature of water *before*	17 °C
temperature of water *after*	57 °C
temperature rise	40 °C
Conclusion: 3 g of butane gas heated 500 cm³ of water up by 40 °C	

Where does the energy come from?

Measurements show that 1 g of butane releases about 50 kJ of energy when it burns. Where does it all come from?

To answer this question we need to write a word equation and then look at the molecules involved.

butane + oxygen → carbon dioxide + water

Here is a butane molecule and an oxygen molecule.

To make carbon dioxide and water from these we must first break all the bonds holding the atoms together. **It takes energy to break bonds**.

Once the bonds have been broken the atoms can form new bonds. **Making bonds releases some energy**.

It takes **about** 150 kJ to break the bonds when burning 1 g of butane. When the new bonds are formed to make carbon dioxide and water about 200 kJ are released. Overall the burning of 1 g of butane gives about 50 kJ of energy.

This energy has not been 'made' in the reaction. It was already there, stored in the bonds of the butane molecules. **When fuels burn, some of the energy stored in their chemical bonds is released as heat**.

Putting out fires

Our lives depend on burning fuels. We work and live in places where fuels are burnt for heating. In the home fuels may be burnt for cooking. We travel in cars and buses where fuels are burnt in the engine. However, we must not forget that burning fuels can be dangerous. A fire which is out of control can damage property. Far worse it can cause serious burns or even death.

The fire triangle

Putting out fires

In a fire, chemical reactions are taking place. A fire needs **fuel** to burn and **oxygen** for the fuel to react with. To keep a fire going some **heat** must reach the fuel.

To put out a fire we must remove at least one of these things. The best method will depend on what is burning and where.

This garden fire is out of control. By hosing it down with water the wood becomes too cool to burn. (The energy is used to evaporate the water.)

The hot fat in this pan has caught fire. The heat supply should be turned off and then the pan should be covered with a lid or a damp cloth. This stops the air from reaching the flames. The lid should be left on for several hours so that the fat cools down.

*You must **never** pour water on to burning fat or oil. The water turns to steam and forces hundreds of tiny drops of burning fat or oil to jump from the pan. This spreads the fire and can cause serious burns.*

This fire was caused by an electrical fault. If you throw water on this while the wire is still 'live' you could get a serious shock. Turn the mains switch off before putting out the fire. Otherwise use sand or a 'dry' extinguisher.

Burning petrol like this can be put out by covering it with a thick layer of sand or earth. Water must not be used because it spreads the burning petrol.

Warning!

If you discover a **small** fire try to put it out. If the fire is too big then leave the room. Close the door behind you to reduce the amount of air getting to the fire. Then get help as quickly as you can.

Fire extinguishers

There are many types of fire extinguishers. Some of the older types used a chemical reaction to force a strong jet of watery liquid and carbon dioxide from the extinguisher. This worked very well on some types of fire, but it could not be used on electrical fires in case the user got an electric shock through the water. The liquid forced out of the extinguisher was also acidic. This caused damage if it was squirted on to curtains or upholstery.

Many modern fire extinguishers use carbon dioxide. It does not burn and is clean and safe to use.

Carbon dioxide extinguisher

High pressure CO_2 extinguisher

This type of extinguisher has a cylinder which contains carbon dixoide gas under pressure. When the trigger is pressed the gas rushes out and smothers the fire. This type of extinguisher is not much use out of doors. The wind blows the carbon dioxide away and lets the oxygen get to the fuel again. It is however excellent for use in the home. It can be small and easy to carry, and it is very easy to use. It does not give out any harmful gases and does not damage furniture. There is also no danger if it is used on electrical fires.

Foam as used at airports

Carbon dioxide foam extinguisher

In this extinguisher chemicals react to make a foam. The foam is a mass of bubbles made from an non-inflammable liquid filled with carbon dioxide gas. The foam stops oxygen getting to the fire. Because the CO_2 is inside the bubbles it does not blow away and so the extinguisher can be used out of doors. Large airports have fire engines which can pump huge quantities of foam over aircraft fires.

Dry powder extinguisher

In this extinguisher high pressure carbon dioxide gas is used to blow a fine, non-inflammable powder over the fire. The carbon dioxide gas and the powder help to smother the flames. This type of extinguisher is safe to use on electrical fires.

First aid treatment for minor burns

Minor burns (like those caused by picking up hot test tubes in the laboratory!) should be treated by holding the burnt area of skin under cold, running water **for at least 10 minutes**. The burn can then be covered with a clean dressing to stop infection. Antiseptic cream can also be used.

Activities

1   Draw a large plan of the school science laboratory. Mark on the things which would be of use in a fire. You should mark fire alarms, smoke detectors, fire exits, fire blankets, fire extinguishers, sand buckets, the main electricity switch, and the main gas tap.

2   Find out about the fire extinguishers in your school. What type are they? How are they used? When were they last checked by the fire officer? How does the fire officer check an extinguisher without setting it off?

3   Design a questionnaire and carry out a survey of your friends and relatives to find out what they know about fire precautions. You could include questions like these:
   a) What would you do *first* if someone's clothes were on fire?
   b) Imagine that you have burnt your hand on the steam from a kettle. What would you do to treat the burn?
   c) Do you have any of the following in your house?
   i) a fire extinguisher   ii) a fire blanket
   iii) a smoke detector.

How long will fossil fuels last?

Over 90 per cent of all the energy used in the United Kingdom comes from fossil fuels – coal, oil, and natural gas. The supplies of these fuels are limited and we are using them up at an alarming rate.

The chart shows how long our fossil fuels will last. Of course we may find some more oil and gas but it would not make much difference to these predictions. You can see that oil and gas are likely to run out in your lifetime!

fuel	known supplies	when is it likely to run out?
natural gas	about 50 years	2035
oil	about 70 years	2055
coal	about 300 years	2285

The fact that we are running out of fossil fuels means that we must plan for the future. There are three things which we can do:

- make the best possible use of the energy we get from our fossil fuels now
- find alternative fuels to coal, oil, and gas
- find new sources of energy where we do not need to burn fuels.

Conservation

Energy conservation means making the best use of our energy supplies. It is important that we do not waste fuel in the home, in our cars, or in our industries.

In the home most of the energy is used to heat rooms. When this energy escapes it goes to heat up the atmosphere outside. It is spread out among many air particles and so is wasted.

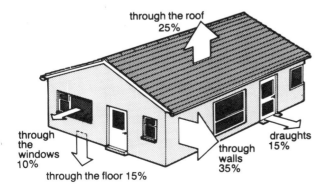

Heat loss from a house

Loft insulation

About 25 per cent (one quarter) of the energy used to heat a house escapes through the roof. This means that for every £100 spent on fuel, £25 is used to heat the air outside! This waste can be reduced by insulating the roof. This is done by covering the floor of the loft (roof space) with a layer of material which it is difficult for heat to get through. For example a thick layer of fibre-glass or mineral wool can be used. The fibres of the insulating material trap lots of air between them. Air is a poor conductor of heat and so not much energy escapes.

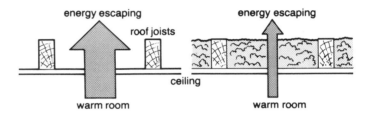

Roof without insulation   *Roof with insulation*

Loft insulation is not very difficult to install and the material is relatively cheap. The money spent on insulating the roof of a house can be saved on reduced fuel bills in about two years.

Cavity wall insulation

About one-third of the heat in our homes escapes through the walls. To prevent this modern houses are built with cavity walls. These have two layers of brick with an air space between. The air acts as a good insulator and so less heat is lost than through a solid wall.

The cavity works even better if it is filled with an insulating material. This traps the air and stops it carrying away heat by convection.

For a new house blocks of mineral fibre can be put into the cavity as the walls are being built. If the house is already built, special insulating foam can be pumped into the cavity through holes drilled in the wall. This is a job which can only be done by specialists. It is expensive so it is likely to take over five years before you save the cost in reduced fuel bills.

Windows

Most buildings have single sheets of glass in the window frames. Heat can be conducted through the glass and lost to the air outside. To prevent this a second sheet of glass can be used in the window. This is called **double glazing**.

Sometimes a second window is added to one which is already there. This is called secondary glazing. The layer of air trapped between the glass acts as a good insulator and so reduces the amount of heat lost.

It is also possible to fit double-glazed window units which have been specially made in a factory. These have two sheets of glass but the air between them is pumped out to leave a vacuum. This reduces heat loss even more.

Special double-glazing units are very expensive and may have to be fitted by experts. It can take up to thirty years to save the cost of double glazing on your fuel bills!

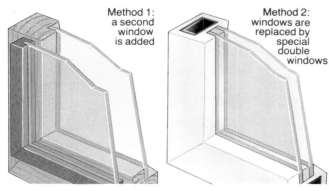

Method 1: a second window is added

Method 2: windows are replaced by special double windows

Double glazing

Other ways of cutting fuel bills

Draught excluders

Draughts let about 15 per cent of the energy escape from our homes. These can often be stopped by fixing strips of plastic or foam around the edges of doors and windows.

Strips which brush against the floor can be fixed along the bottom of doors. This lets them open and close but stops cold air from blowing in under the door. Draught excluders like these are very cheap and easy to fit. They make the house more comfortable and their cost can be recovered very quickly.

Hot water cylinder jackets

In houses, hot water is usually stored in a copper cylinder. Copper is a very good conductor of heat so the cylinder loses energy quickly. The immersion heater or boiler then has to heat the water again. Covering the cylinder with a thick 'jacket' filled with insulating material slows down the heat loss and so saves fuel. The cost of insulating the hot water tank can be recovered in a few months.

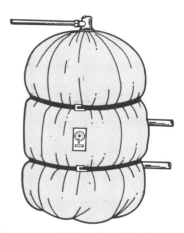

Insulating the hot water cylinder saves a lot of energy

Activities

1 Find out about energy conservation in your home.
   a) If you have a loft is it insulated? What type of insulation is it? How thick is the layer of insulation? Is the cold water tank lagged?
   b) Have you got double glazing? If you have is it secondary glazing or factory-made units? Try to find out how much it costs to have double-glazed windows fitted.
   c) Have you got cavity walls? (This may be difficult to check but you can do it by measuring the thickness of an outside wall. A solid brick wall will be about 25 cm thick. A cavity wall will be about 30 cm thick.)
   If you have got cavity walls are they filled with insulating material?
   d) Are your doors and windows draughty? Are they fitted with draught excluders? Find out how much a draught excluder costs for   i) the bottom of a door   ii) a letter box.
   e) Is your hot water tank lagged? Does the jacket fit well or are there gaps in it?

Other sources of energy

The way we use fuels in our homes has changed very much over the past hundred years or so. Originally candles and oil lamps were used for lighting. They were dirty, dangerous and did not give much light. Eventually coal gas was piped into houses. Gas lamps lit houses and streets. This was much more convenient than oil but still it did not give a good light. Finally, electricity took the place of gas and now electric light bulbs and fluorescent tubes seem to be everywhere.

Electricity has changed our lives because it can be used for lighting, heating, and to operate motors inside domestic appliances like vacuum cleaners and hair driers. Electricity is not a fuel. We produce nearly all our electricity by burning fossil fuels.

Hot water geyser

Fossil fuel to electricity

The fuel, usually oil or coal, is burnt in boilers. The energy heats water and turns it to steam. The hot steam is then piped to steam turbines. A steam turbine is like a fan. When the steam flows through it the blades of the turbine turn. This drives an electrical generator.

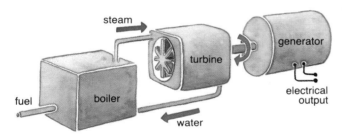

The electricity produced can be stepped up to very high voltages by a transformer and then sent all over the country using the National Grid system.

Because our supplies of coal and oil are running out and because of the pollution caused by burning fossil fuels, other ways of making electricity must be found.

Nuclear fuels

A thermal power station uses coal to heat water. This can also be done using the energy released in a nuclear reactor using uranium as a fuel (see page 105).

Uranium is an alternative fuel and it has been used in the United Kingdom for many years. There is disagreement over whether we should be using nuclear fuels or not. Some of the arguments for and against can be found in Chapter 5.

Geothermal energy

The inside of the Earth is very hot. No one knows exactly how this energy is generated (except that it is a radioactive process) but we do know that there are hot rocks just below the surface. Volcanoes and hot water geysers give us dramatic evidence for this.

In the future it may be possible to use the hot rocks commercially in the UK to turn water into steam which could then be used to turn turbines and generate electricity. Test drillings are already being made to find the best way of using geothermal energy.

Questions

1 a) Describe two advantages of gas lighting over the use of candles.
   b) Describe two advantages of electric light bulbs over gas lights.

2 Explain why electricity should not be described as a 'fuel'.

3 Explain why uranium should not be described as a 'fossil fuel'.

4 Design a method for generating electricity from the energy trapped in hot rocks below the ground. You should illustrate your answer with a clear, labelled diagram of your method.

Describe any problems which you think might occur with your method.

Renewable energy sources

Energy sources which do not use fuels dug from the Earth are called **alternative** energy sources. They use natural sources such as the sun, the tides, and the wind. Because these things will be able to supply energy for millions of years they are called renewable energy sources.

Because we have been able to get energy easily by burning coal and oil we do not have much experience of using these alternative sources of energy. There are many problems to be overcome but scientists and engineers have started to develop new ways of using renewable energy sources.

Traditional windmill

Wind power

The old windmills of England and the Netherlands show that we can use energy from the wind. These old mills were used to grind corn but when the wind did not blow the work had to stop. The fact that the wind does not blow steadily is still a problem. Windmills or, more correctly, wind turbines can be used to drive electrical generators but storing the energy is not easy. One suggestion is that we should build a large number of huge wind turbines off the coast of Scotland. The electricity could be fed into the National Grid. When the wind was blowing, other types of power stations could produce less electricity and so burn less fuel.

Modern wind turbine

Hydro-power

Around 200 years ago mills and factories were built near to fast flowing rivers. The moving water was used to turn a water wheel which then drove machinery. Old water wheels can be found throughout the counry but they are rarely used now. However, we can use flowing water to drive turbines in hydro-electric power stations.

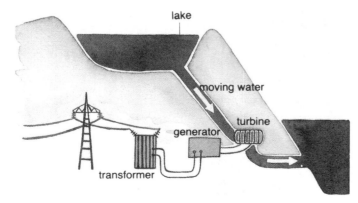

Hydro-electric power station

Hydro-electric power stations have proved very successful in some parts of the world such as Scandinavia and North America but they are expensive to build. It is often necessary to change the surrounding countryside by building dams and diverting rivers. This may flood farmland and people may have to be moved from their homes. Some hydro-electric projects are facing problems because mud carried by the rivers is building up behind the dams and may block the turbines.

Questions

1 What is a 'renewable' source of energy? Give one example.

2 What is the main problem with trying to use wind power?

3 Why do you think that the north coast of Scotland has been suggested as a good site for wind turbines?

4 What is meant by 'hydro-power'? Write down three other words that use *hydro* to mean water.

5 What would you look for if you were trying to find a site to build a hydro-electric power station?

6 Suggest two reasons why a hydro-electric power station may not be a good idea in Ethiopia.

More alternative energy sources

Solar energy

The Earth gets more than enough energy from the sun to meet all its energy needs. If we can trap and use this energy we will not need to burn fuels. Scientists and engineers are trying to find the best way to do this.

Solar panels

A panel, like this, on the roof of a house uses energy from the sun to heat water. It is called a solar panel.

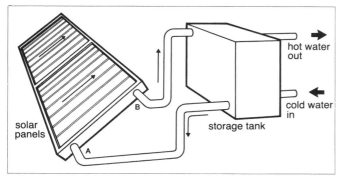

Heating water using a solar panel

Solar panels are very useful in parts of the world where the sun shines nearly every day. They are less useful in the United Kingdom because our sunshine is more variable. We also get most of our sunshine in the summer when we need less energy!

Solar cells

The calculator shown here is solar powered. The solar cells use the energy in light to produce a very small amount of electricity. This is enough to make the calculator work and so no batteries are needed. Cells like these are used to generate electricity for satellites in space.

Power from the sea

This sea is constantly in motion. Research is being carried out to see how we can use the tides and waves to produce electricity.

Tidal barrage

A tidal power station is built on a long barrier called a tidal barrage. This is a special barrier built across the mouth of a tidal river (the **estuary**).

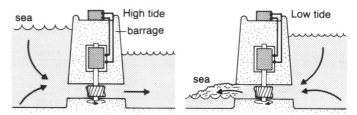

A tidal barrage. Water flows from the sea through the turbines in the barrage. Electricity is produced and the water is trapped. Water is released to flow through the turbines. These drive generators to produce electricity.

Tidal barrages are very expensive to build and, because they change the flow of the river, they can affect the environment. This may mean that sea birds and other animals can no longer live in the estuary.

Wave power

There are plans to use sea waves to generate electricity. One idea is to build huge floating generators. These would move up and down with the waves. This movement would drive the generators.

Another idea is to let the sea waves move up and down inside large tubes. As the wave moves up the air in the tube is compressed. The compressed air can then be used to turn a turbine connected to a generator.

These ideas are only experimental. Models have been made but it will be many years before full-size wave-power generators are built.

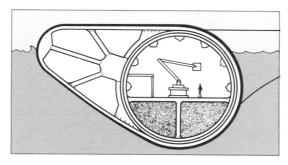

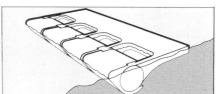

Salter's ducks are wave-powered generators. As they nod up and down on the waves (like ducks!) they generate electricity.

Energy from living things

Green plants have been trapping energy from the sun for millions of years. They use the energy to make sugars from carbon dioxide and water. This is called **photosynthesis** (see Chapter 6). There are ways of releasing this energy so we can think of plants as sources of renewable energy.

Fermentation

The sugars made by plants can be changed into alcohol by using yeast. The method for doing this is called **fermentation** (see page 178). We use this to make alcoholic drinks like wine, but alcohol will burn and so it can be used as a fuel.

In some parts of the world, for example in Brazil and Zimbabwe, special crops like sugar cane are grown to make alcohol. It grows very quickly and the sugar it contains is very easy to ferment. The alcohol is then mixed with petrol and used by cars and lorries.

Energy from biological waste

Decaying plants can give energy. This can be easily shown by filling a vacuum flask with fresh grass clippings. A thermometer is then fixed into the neck of the flask so that its end is resting in the centre of the grass. After a few days the temperature is much higher. The energy has been released by the microbes which have started to digest the grass.

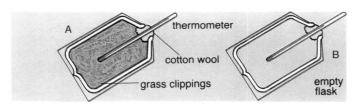

(Flask B is set up as a control. The only thing missing is grass. The control will show whether the temperature rise is due to the grass or something else.)

Using animal waste

In some parts of the world animal dung (faeces) is used as a fuel. It is dried in the sun and then burnt. Unfortunately this means that it is not being used as a fertilizer to improve the soil.

When animal and plant waste decays under the right conditions, methane gas is given off. In a modern sewage system this can be collected and used as a fuel.

Some villages in developing countries now have **biogas generators**. Animal waste is put into a tank with a lid. As microbes digest the waste, methane gas is given off. This can be used for boiling water and for cooking.

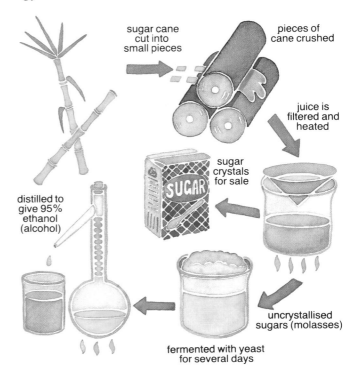

Making ethanol from sugar cane

Activities

Many gardeners build heaps of waste plant materials to make compost. The old plants rot as microbes and other organisms digest them. This produces enough heat to kill off any weed seeds on the compost heap.

Design an experiment to show that the inside of a compost heap gets hot. Your method should allow you to investigate the middle of the heap without letting in too much cold air from the outside. You may want to use plastic bottles, lengths of hosepipe, water, and thermometers.

Questions

1 Where does the energy in green plants come from?

2 Plants make sugars by photosynthesis. What gas do they use?

3 Why can we think of plants as a 'renewable' energy source?

4 Why is sugar cane a good crop to grow for making alcohol in countries like Brazil?

Questions

1 Coal is an impure form of carbon. One of the impurities is sulphur. Carbon and sulphur both burn in oxygen.

a) Write a word equation for the combustion (burning) of carbon in oxygen.

b) How could you show that the gas given off when carbon burns is carbon dioxide?

c) Write a word equation for the combustion of sulphur in oxygen.

d) How could you show that the gas given off when sulphur burns is acidic?

e) Explain why coal-fired stations in the north of England can damage trees in Norway.

2 The diagram shows a tower used for the fractional distillation of oil. The table lists some fractions and their boiling points.

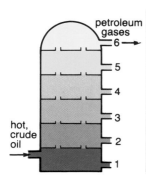

Fraction	boiling point
bitumen	350 °C
lubricating oil	270 °C
diesel oil	200 °C
paraffin	150 °C
petrol	50 °C
petroleum gases	less than 20 °C

a) Which fraction would be taken out at the bottom of the tower (level 1)?

b) What could this fraction be used for?

c) At what level could you collect petrol?

d) Which fraction is used for jet engines?

e) Name *two* things other than fuels which can be made from the chemicals extracted from crude oil.

3 Explain briefly how you would deal with the following fires in the home.

a) A young friend is playing with matches and sets his clothes on fire.

b) A faulty electric blanket sets the bed on fire. (You are not in it!)

c) You smell smoke and discover a large fire in an empty room.

d) You are frying eggs when the fat in the frying pan catches fire.

4 The White family live in the house shown below. They spend £600 each year to heat the house. The

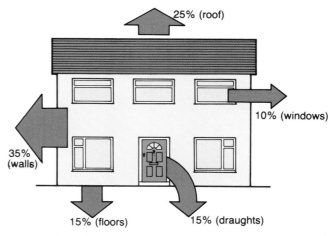

arrows show how their house loses energy to the outside.

a) The Whites decided to insulate their roof. The insulation cost £150 and they found that their fuel bills were £50 cheaper each year.

How long will it take for the fuel savings to pay for the cost of insulating the roof?

b) Pleased with the roof insulation, the Whites had the cavity walls of their house filled with foam. This cost £900 and their fuel bills went down by £150 per year.

  i) What is a cavity wall?

  ii) How does the foam reduce heat loss?

  iii) How long will it take for the fuel savings to pay for the cost of the cavity wall insulation?

c) Finally the Whites had all their old windows replaced by double-glazing units. This cost £3000. Their fuel bills went down by £50 per year.

  i) What is double-glazing?

  ii) How long will it take for the fuel savings to pay for the double-glazing?

  iii) The White's house is near a busy road. Suggest one reason why they are very happy with their double-glazing.

5 a) What is meant by the term 'fossil fuel'?

b) Explain why scientists are trying to find alternative energy sources now.

c) Describe one alternative source of energy that may be producing electricity for your home in the year 2055.

5 Radioactivity

What is radioactivity?
Where does radioactivity come from?
How can we use it?
Is it safe?

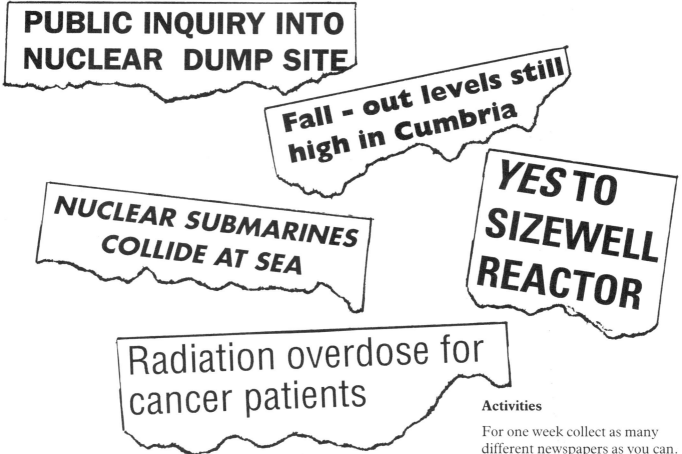

PUBLIC INQUIRY INTO NUCLEAR DUMP SITE

Fall - out levels still high in Cumbria

NUCLEAR SUBMARINES COLLIDE AT SEA

YES TO SIZEWELL REACTOR

Radiation overdose for cancer patients

Headlines like these are very common. Nearly every day the newspapers have articles about nuclear weapons, disposal of waste from power stations, or leaks from nuclear processing plants. The public is right to be worried about these things but it would be wrong to believe that all radioactivity is bad. Scientists have been able to make good use of nuclear energy.

Industry and medicine use radioactivity in ways which help society. Of course there are also real dangers, so scientists must use their knowledge of nuclear reactions to keep the risks down to the lowest possible level.

All people should be able to join in the public debate about the use of nuclear power. By studying radioactivity we can support our arguments for or against it with scientific evidence. We will also be able to decide whether politicians and other public figures are giving a fair picture of the situation.

Activities

For one week collect as many different newspapers as you can. You can also collect news and scientific magazines. Cut out all the articles which have anything to do with nuclear energy, nuclear weapons or radioactivity. Use these to make a poster for display. Answer these questions:

1 How many articles were about the *bad* effects of nuclear energy? How many articles were about the *good* effects?

2 What sort of things do people worry about when thinking of nuclear energy? Ask friends and relatives.

Inside the atom

Elements are made up of atoms. An atom is so small that it is almost impossible to imagine. A single atom measures about 1/10 000 000 mm across. The full stop at the end of this sentence contains more than ten million million atoms.

In diagrams, an atom is usually drawn as a solid ball. In fact there are even smaller particles inside. The easiest ones to remove are the **electrons**. These are found in the outer part of the atom. Electrons have a negative electrical charge but have very little mass.

The electrons move around in a space which takes most of the atom. The other particles are packed together in a very small **nucleus** at the centre of the atom. The nucleus is about 1/100 000 000 000 mm across!

We find two types of particles inside the nucleus: **protons** and **neutrons**. Protons have a positive charge. Because positive charges attract negative charges the protons attract the electrons. Each proton is about 2000 times heavier than an electron.

Neutrons do not have any charge: they are neutral. They are also much heavier than electrons. In fact the mass of a neutron is almost exactly the same as the mass of a proton.

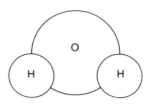

This diagram represents two atoms of hydrogen joined to one atom of oxygen. This makes one molecule of water. (Real atoms are not solid balls.)

particle	where found	electrical charge	mass (proton = 1)
electron	outside nucleus	−1	$\frac{1}{1840}$ (almost zero)
proton	in the nucleus	+1	1
neutron	in the nucleus	0	1

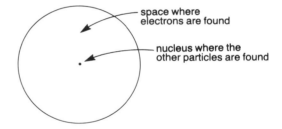

Diagram of one atom – not to scale: the nucleus should be drawn even smaller. If it was the size shown the atom would be about 10 m across!

How many protons?

Each element has a fixed number of protons in the nucleus of its atoms. This is the element's **atomic number**. For example, hydrogen has atomic number 1. This means that each hydrogen nucleus has one proton. (A hydrogen atom has one electron to 'balance' the one proton in the nucleus.)

Carbon has atomic number 6. All carbon atoms contain six protons and six electrons.

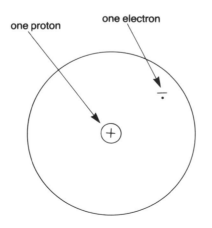

Hydrogen atom

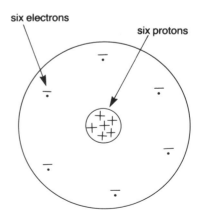

Carbon atom

Neutrons and isotopes

How many neutrons?

Neutrons and protons have mass. The more neutrons there are in an atom the heavier it gets. This changes the density and other physical properites of the element. However, neutrons do not have any electrical charge and so they do not affect the chemical reactions of an element. To find out how many neutrons there are in a nucleus we need to know the atom's **mass number**. This tells us how many particles (protons and neutrons) there are in the nucleus. A better term is **nucleon number**. Some examples are given in the table.

element	atomic number (number of protons)	nucleon number (number of protons + neutrons)	nucleon number − atomic number = number of neutrons
hydrogen	1	1	0
carbon	6	12	6
sodium	11	23	12
gold	79	197	118

Isotopes

An element always has the same number of protons in its atoms but the number of neutrons is not fixed. For example, every hydrogen atom has one proton. Most do not have any neutrons but some have one and others have two! This means that the element hydrogen has three different types of atom. These are called **isotopes**. Hydrogen has three isotopes.

element	atomic number	nucleon number	number of protons	number of neutrons	diagram
hydrogen (normal)	1	1	1	0	
hydrogen (deuterium)	1	2	1	1	
hydrogen (tritium)	1	3	1	2	

nucleon number (or mass number) shows that there are 23 particles in the nucleus; 11 protons and 12 neutrons

symbol for the element sodium

Na

atomic number shows that there are 11 protons in the nucleus

Using symbols

Information about the structure of an atom can be given in a symbol. The symbol above right tells us about a sodium atom.

It is easy to compare isotopes using these symbols.

symbol	$^{12}_{6}C$	$^{14}_{6}C$
name of isotope	'carbon-12'	'carbon-14'
protons	6	6
neutrons	6	8

Questions

1 Draw diagrams to show the structure of the following atoms:

$^{4}_{2}He$ (helium) $^{7}_{3}Li$ (lithium) $^{14}_{7}N$ (nitrogen) $^{16}_{8}O$ (oxygen)

2 Write down the number of protons and the number of neutrons in these atoms:

$^{207}_{82}Pb$ (lead) $^{238}_{92}U$ (uranium) $^{242}_{94}Pu$ (plutonium)

The discovery of radioactivity

In 1896, a French scientist called Henri Becquerel was investigating X-rays which had been discovered just one year earlier. He placed some photographic plates, wrapped in paper, near some compounds containing the element uranium. When he developed the plates he found that they were 'fogged' just as if they had been in the light or had been exposed to X-rays. Becquerel decided that the uranium was giving out a type of ray never studied before. This was called radioactivity; a word we still use.

During the next few years scientists investigated this new radiation. Marie and Pierre Curie found other radioactive elements such as radium and polonium. In 1899 a New Zealander, Ernest Rutherford, found that the radiation from uranium was in fact two different types: **alpha** (α) and **beta** (β). In the same year a third type of radiation, **gamma** (γ), was discovered by a French scientist, Paul Villard.

All this early work was very exciting but real understanding of radioactivity came in 1902 when Rutherford and Soddy suggested that radioactivity results when the nucleus of an atom changes. This was a remarkable idea because it meant that elements were changing into other elements; something which was not thought possible at the time.

Marie and Pierre Curie

Detecting radiation

Photographic film

Becquerel first detected radiation from uranium using photographic plates. This method is slow because the 'photograph' has to be developed. However, it is still used. Workers who use radioactive materials wear special badges with a piece of film sealed inside. When the film is developed the 'fogginess' of the film shows how much radiation has reached the worker. This is an important safety check.

Radiation monitoring badge

Geiger counter

Radiation carries energy so it can 'knock' electrons out of atoms. This leaves charged ions. In a **Geiger–Müller tube** (GM tube) electrons released by radiation are attracted to a positive electrode (anode). This gives tiny pulses of current which can then be counted on a special counter. The diagram shows the parts of a GM tube. The amplifier can be connected to a counter which will count every pulse of energy. To find out what the radiation level is, you need to find the number of 'counts' in one minute or in one second. This is the **count rate**.

Sometimes the GM tube is connected to a rate meter. The meter has a scale which shows the count rate in counts per second. The higher the radiation level the more counts per second.

Geiger counters are often fitted with a loudspeaker. This gives a click every time a pulse of energy enters the GM tube. With a weak source near the detector, the loudspeaker gives a slow, irregular series of clicks. When a strong source is nearby the clicks get closer and closer together. Geiger counters like this are very useful for monitoring radiation levels where people are working because the noise acts as a warning.

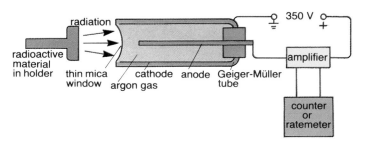

A Geiger counter

Background radiation

If a Geiger counter with a loudspeaker is set up far away from any laboratory radioactive sources, a slow, irregular clicking is heard. This tells us that there is some radiation about even when we are not using radioactive materials. This radiation is always around us and so is called the **background radiation**.

The level of background radiation is usually very low but it does vary from place to place. Most of the background radiation comes from rocks in the Earth's crust. Some comes from cosmic rays from outer space and some comes from the radioactive elements inside our bodies! Even some of the gases we breathe are radioactive.

The dose of radiation you get during a year will depend on where you live, whether you travel by air, and whether you have had medical treatment. The numbers in the chart are about average for the United Kingdom.

Using radioactive materials safely

Radioactive materials must always be handled with care. They must be kept in boxes lined with lead and should be locked away after use.

Schools usually use small, weak radioactive sources which are mounted in special holders. When using these, the teacher will use tweezers and will make sure that the source is not pointed towards other people.

If these simple steps are taken, radioactivity experiments can be safely carried out in the school laboratory.

In industries where the level of radiation is high, workers have to wear special clothes lined with lead. After use, these are sealed in drums and disposed of. The workers also wear photographic badges to check that they have not received a dangerously high dose of radiation (see page 102).

Some radioactive sources are too dangerous to handle. These can be used inside a lead-lined cabinet. Scientists then use remote-controlled arms to move the source and other bits of apparatus. Video cameras mounted on small, remote-controlled trolleys can be used to inspect the inside of nuclear reactors. These robots can go into places where the radiation level is high enough to kill any human.

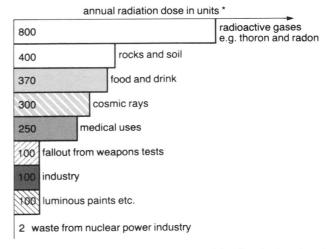

annual radiation dose in units *

800	radioactive gases e.g. thoron and radon
400	rocks and soil
370	food and drink
300	cosmic rays
250	medical uses
100	fallout from weapons tests
100	industry
100	luminous paints etc.
2	waste from nuclear power industry

* 1 unit = 1 microsievert

The background radiation absorbed by a person in one year

Questions

1 How many units does the person absorb in one year?

2 How many of these units are from 'natural' sources?

3 How many of these units are from human activities?

4 Roughly how many units of radiation will a person absorb in his or her lifetime?

Highly radioactive materials are handled inside shielded cabinets.

Types of radiation

The diagram opposite shows the basic apparatus used for radioactivity experiments in schools. The radioactive source is inside a special holder which can be moved towards or away from the detector.

There are three types of nuclear radiation: alpha, beta, and gamma. Radioactive materials give out one or more of these types.

Each type can be identified by the ease with which it passes through materials (penetrating power) and by the way it behaves in a magnetic field.

Penetrating power

Different materials are placed between the source and the detector. We can use paper, cardboard, thin aluminium sheets, and pieces of lead. By observing what happens to the count rate when the absorber is put in the way we can find what type of radiation is being emitted.

Deflecting radiation with a magnet

A counter can be set up to detect a beam of beta radiation from a strontium-90 source. When a strong magnet is placed with its poles on either side of the beta radiation the count rate immediately drops showing that less radiation is getting to the detector. It has been deflected by the magnet. The direction shows us that the radiation is carrying negative electrical charge.

Careful measurements show that beta radiation is a stream of high energy electrons.

Similar investigations show that alpha radiation is deflected (less) in the opposite direction.

It can be shown that alpha radiation is a stream of helium nuclei. Each alpha particle contains two protons and two neutrons.

When the experiment is repeated with a gamma source, no deflection can be found. **Gamma radiation cannot be deflected by a magnetic field because it is an electro-magnetic wave, like light**.

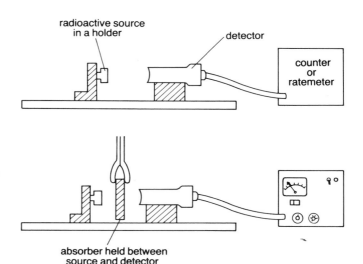

Investigating penetrating power

summary of results			
type of radiation	penetrating power	magnetic deflection	what is it?
alpha (α)	– will not go through thin card – stopped by a few centimetres of air	small deflection	positively charged helium nucleus (two protons and two neutrons)
beta (β)	– penetrates thin card and foil – stopped by a few millimetres of aluminium	large deflection	high energy electrons (negatively charged)
gamma (γ)	– very penetrating – stopped by several centimetres of lead or several metres of concrete!	no deflection	electro-magnetic wave. Very short wavelength; very high energy.

What causes radioactivity?

Radioactivity is caused when a nucleus of an atom breaks up. We say that the nucleus is **unstable**. The particles in the nucleus stay together for a time and then suddenly fly apart without warning. The particles and energy given out are detected as radiation.

Heavy elements like uranium, radium and plutonium, have no stable isotopes. It is as if they have too many particles in their nuclei.

Many elements have several different isotopes. Some of these may be stable. These are not radioactive. Other isotopes may be unstable. These are called radioisotopes. For example, hydrogen has three isotopes; ^1_1H ^2_1H and ^3_1H. Normal hydrogen (^1_1H) and deuterium (^2_1H) are stable but tritium (^3_1H) is radioactive.

For light elements like carbon and hydrogen the nucleus seems to become unstable when the number of neutrons is bigger than the number of protons.

isotope	number of protons	number of neutrons	stable or radioactive
^1_1H	1	0	stable
^2_1H	1	1	stable
^3_1H	1	2	radioactive
$^{12}_6C$	6	6	stable
$^{14}_6C$	6	8	radioactive
$^{235}_{92}U$	92	143	radioactive

Radioactive decay

When a nucleus gives out an alpha or beta particle, it changes into a different element. We say that the original atom has **decayed**.

In **alpha decay** the 'parent' nucleus loses 2 protons and 2 neutrons as an alpha particle. The atom which forms is 4 'units' lighter and has an atomic number which is two less. The diagram and equation below show the decay of a uranium-238 atom.

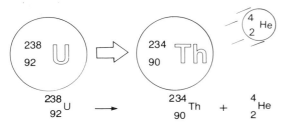

$$^{238}_{92}U \longrightarrow ^{234}_{90}Th + ^4_2He$$

The decay of uranium-238

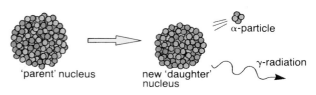

Radioactive decay

In **beta decay** a nucleus gives out an electron. This is surprising because we have already seen that there are no electrons in the nucleus! (See page 95.) It is as if a neutron inside the nucleus has split up into a proton and an electron. The electron is then ejected as a beta particle. The diagram and equation below show the decay of a tritium (hydrogen-3) atom.

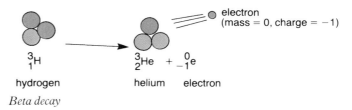

$$^3_1H \longrightarrow ^3_2He + ^0_{-1}e$$

hydrogen   helium   electron

Beta decay

Worked example

Question Write down the equation for the alpha decay of americium-241. (The atomic number of americium is 95.)

Answer

$^{241}_{95}Am$ has 95 protons and a mass of 241 units.

When it decays it will lose four mass units. Two of these will be protons and two will be neutrons.

$$^{241}_{95}Am \longrightarrow ^{237}_{93}X + ^4_2He$$

We do not know what the new element is so we call it X. We do know that it has 93 protons (95 − 2). We also know that it has a mass of 237 (241 − 4).

Using a table of atomic numbers we can find that X is actually the element neptunium (Np).

Questions

1 Write nuclear equations for the alpha decay of these isotopes. (You can use the letter X for the new elements.)

$^{234}_{92}U$ $^{226}_{86}Ra$ $^{214}_{84}Po$

2 Write nuclear equations for the beta decay of these isotopes.

$^{234}_{90}Th$ $^{214}_{82}Pb$ $^{210}_{83}Bi$

Half-life

We cannot control radioactive decay but we can measure how fast it takes place. One way of doing this is to take lots of atoms of a radioisotope and to measure how long it takes for half of them to change. This is called the **half-life**.

The half-life is the time taken for half the atoms in a sample of radioactive isotope to decay.

For example, the half-life of cobalt-60 is about 5 years. If we start with 16 g of cobalt-60 it will give out radiation. After 5 years only 8 g of cobalt-60 will be left. The other 8 g will·be made up of different types of atom. These are the decay products. The diagram shows what happens to the cobalt-60 over four half-lives (20 years).

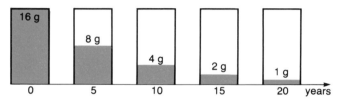

Diagram showing the radioactive decay of cobalt-60 over four half-lives

Each isotope has its own half-life. Some are very unstable and have half-lives of less than a second. Others have very long half-lives of thousands, or even millions of years.

Measuring half-life

Radon-220 is a radioactive gas which emits alpha particles when it decays. The half-life of radon-220 can be measured by seeing how quickly the radiation level falls for a sample of the gas.

The gas is produced in a bottle containing thorium hydroxide. When the bottle is squeezed, some radon-220 is pushed into the container with the GM tube. After a few squeezes the clips are closed so that no more gas can get in. The counter is switched on and the reading taken every 30 seconds. The results from an experiment are plotted on the graph.

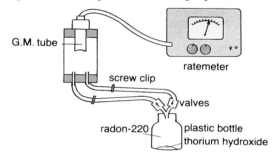

Measuring the half-life of radon-220

The graph shows that the half-life of radon-220 is 52 seconds.

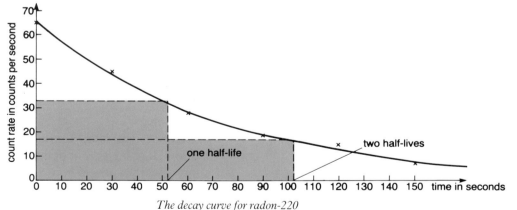

The decay curve for radon-220

Questions

1 Iron-59 has a half-life of 30 days. If you start with 1 g of iron-59 how much will be left after   **a)** 30 days   **b)** 60 days   **c)** 120 days?

2 A sample contains 0.1 g of radium-224. After 7 days the sample contains just 0.025 g of radium-224. What is the half-life of this isotope?

3 Protactinium-234 has a half-life of 70 seconds approximately. A sample of this isotope gave a count rate of 80 counts per second. Draw a graph showing how the count rate varied over the next 300 seconds.

4 Radon-220 emits alpha particles. Explain why it is important to make sure that the plastic bottle does not leak in the experiment described on this page.

Radiation and living things

Radiation from radioactive materials carries energy. If radiation is absorbed by a living cell the energy may kill the cell or change the chemicals inside it. These effects make radioactivity dangerous if proper safety precautions are not taken.

Effect on humans

Alpha particles cannot penetrate the surface of the skin but if an alpha-emitting gas is breathed in, the radiation cannot get out of the body. This makes it very dangerous. Beta particles and gamma radiation can penetrate the skin and can damage cells. The harmful effects are divided into two groups; **early effects** and **late effects**.

Early effects of radiation

Early effects are due to exposure to high levels of radiation. People having medical treatment for cancer, nuclear reactor workers close to accidental leaks, and survivors of a nuclear attack may experience early effects within a few days of exposure. The effects, shown in the table below, depend on the size of the dose received.

dose in units★	early effects
2000	None. This is the average dose received in one year from natural 'background' radiation.
100 000	Possible temporary sterility.
1 000 000	Hair falls out. Skin is red and sore. Sickness.
4 000 000	Half the people receiving this dose will die.
10 000 000	All humans would die.

(★ 1 unit = 1 microsievert)

Late effects

Late effects usually appear as **cancer** and occur several years after exposure.

Cancers seem to be caused when radiation changes the chemicals inside the nucleus of the cell. This does not kill the cell but makes it divide and grow as if it was out of control. Sometimes the cells divide so quickly that a lump or tumour forms.

Studying the effects of radiation

Experiments must not be carried out on humans but plants, insects, and small animals are used for research. These tests help us to develop uses for radiation.

Effect on plants

Seeds taken from a healthy plant are given measured doses of gamma radiation. The seeds are then planted in moist soil to germinate (see page 229). After the plants have been growing for several days they are studied. The photograph shows the results for seeds given increasing amounts of radiation.

The effect of radiation on living things. The plant on the left was grown from normal seed (control). The other plants were treated with increasing doses of gamma radiation.

The seeds which had low doses have grown well. They have long roots and large leaves. Seeds which had higher doses are smaller. Some seeds may not germinate at all.

The radiation has penetrated the cells inside the seeds. A very high dose destroys the cell but lower doses change the cells so that they cannot produce DNA. and so the cells cannot divide and grow.

Mutation

Sometimes irradiated seeds grow into plants which do not look like the parent plants. Some may be very short (dwarfs), others may have strange leaf patterns, and others may not have any colour (albino). These variations are called **mutants**. They look strange because the radiation has changed the genes (see page 233) of the cells.

Some of the changes are hereditary and can be passed on to the next generation.

Most of the mutations will be useless and will be thrown away but one or two may be 'better' than the parent plants. The useful mutations can be bred to give farmers better crops.

Radiation in medicine

Sterilizing medical instruments

Bacteria are small, living organisms found almost everywhere around us. They can cause disease and infection so great care must be taken in hospitals to use surgical instruments which have no bacteria on them (they are **sterilized**). Glassware and metal instruments are put into boiling water or steam. The very high temperature kills the bacteria. This method could not be used with plastic bowls or syringes because they would melt. These, together with bandages and dressings, can be sterilized using a large dose of gamma radiation to kill the bacteria.

Treatment of cancer

Cancer cells can be killed using radiation. If the cancer cells are deep inside the body, gamma rays from a cobalt-60 source are used to destroy them. The intensity of the radiation and the direction of the rays have to be carefully calculated. Several treatments are usually necessary.

Patients being treated with radiotherapy may lose their hair temporarily. They may also suffer from radiation sickness. However, the treatment is often completely successful and cures cancer which would have killed the patient.

The cells of skin cancers are close to the surface and so can be treated with beta radiation. A strontium-90 source is strapped on to the skin above the cancer. The radiation kills the cancer cells but does not penetrate far enough to cause any other damage.

Improved drugs, surgical techniques and radiation treatment mean that more and more cancers can be successfully treated. Today about 90 per cent of skin cancers can be cured.

Studying the thyroid gland

The thyroid gland is found in the throat. It produces a powerful chemical, the thyroid hormone, which controls growth in the body. If the hormone level is too high or too low the person will become ill. Symptoms include loss of weight, bulging eyes, swelling of the neck and trembling hands.

The thyroid gland uses iodine to help make the hormone. By using a solution containing a small amount of radioactive iodine-131, doctors can measure how much is absorbed by the gland. They can then give drugs or use surgery to make the gland work properly.

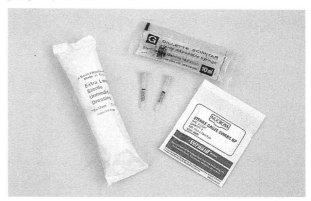

Plastic syringes can be sterilized using gamma radiation.

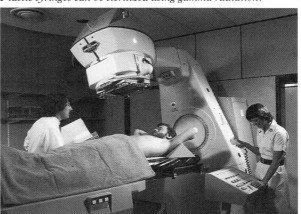

Radiotherapy; curing cancer

Questions

1 Why must gamma radiation be used to treat cancer tumours inside the body?

2 Why does the gamma radiation source have to be positioned very carefully in the treatment of cancer?

3 What metal, do you think, is used to shield the doctors and nurses from the gamma radiation?

4 What 'early effects' of radiation are shown by patients having radiation treatment?

5 After the Chernobyl disaster a lot of radioactive iodine-131 was produced. Doctors in some countries prescribed non-radioactive iodine tablets for children.

    a) What gland could be damaged by radioactive iodine?

    b) Suggest why ordinary iodine tablets were given.

Splitting the atom

In the nuclear reactions we have seen so far the radioactive nucleus ejects an alpha particle or a beta particle. These are very small so the nucleus which is left is almost as big as the original one. For example, when americium-241 decays it gives out an alpha particle (helium nucleus).

$$^{241}_{95}\text{Am} \rightarrow {}^{237}_{93}\text{Np} + {}^{4}_{2}\text{He}$$

americium	neptunium	helium
(mass 241)	(mass 237)	(mass 4)

The neptunium formed has a mass of 237; not much less than 241.

There is another type of reaction where a nucleus can be split roughly in half by collision with a neutron. This is called **nuclear fission**. It can happen in uranium.

Notice that in fission one free neutron goes in but two or three come out. These can go on to cause other uranium nuclei to split giving even more free neutrons. This is called a **chain reaction**.

Uncontrolled chain reactions

In a small piece of uranium the chain reaction never really gets started because many neutrons escape from the surface.

In a large piece of uranium not enough neutrons can escape and so the chain reaction gets out of control. In this process a huge amount of energy is released in a very short time and this may result in an explosion.

The energy produced in a nuclear explosion is huge. **A single modern nuclear warhead can give an explosion bigger than one million tonnes of chemical explosive**. In addition, a nuclear explosion produces radiation which would make the environment unsafe for living things for many years.

Activities

1 Find out about Hiroshima and Nagasaki.

2 Find out what is meant by 'nuclear deterrent'.

3 Find out what the letters CND stand for? What are the aims of CND?

4 How do countries test nuclear weapons?

5 What is the Strategic Defence Initiative (SDI)?

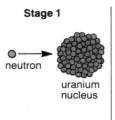
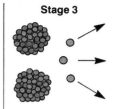

Stage 1	Stage 2	Stage 3

neutron

uranium nucleus

A neutron is absorbed by a uranium nucleus

The nucleus with the extra neutron starts to split

The nucleus splits into two, roughly equal parts. These are the fission products. Two or three high energy neutrons are ejected.

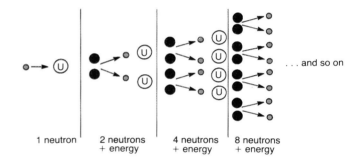

... and so on

1 neutron   2 neutrons + energy   4 neutrons + energy   8 neutrons + energy

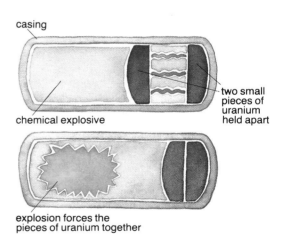

casing

chemical explosive

two small pieces of uranium held apart

explosion forces the pieces of uranium together

One type of atom bomb

Controlling nuclear fission

In 1938 a scientist called Otto Frisch showed that splitting uranium atoms would release heat. Four years later, Enrico Fermi showed how to control this release of energy. His work made **nuclear reactors** possible. Now many countries use nuclear power stations to produce electricity.

Most nuclear power stations use uranium as a fuel. The uranium atoms undergo fission and release heat. The diagram below shows how the heat is used to produce electricity.

There are several different types of nuclear reactor. One type is shown in the diagram opposite.

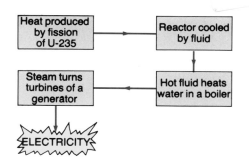

How does it work?

The fuel rods
The uranium fuel is packed in stainless steel tubes. The uranium is purified and has been 'enriched'. This means that it has been treated to increase the number of uranium-235 atoms. These are the atoms which undergo fission giving out energy and neutrons to keep the chain reaction going.

The graphite moderator
Inside the reactor there are plenty of fast-moving neutrons produced by radioactive decay. These can start a chain reaction but if they are moving too quickly they pass the uranium nuclei without being trapped. The graphite inside the reactor slows them down. This means that more neutrons will be absorbed by the nuclei and so more nuclei will split and give out energy.

The control rods
Without control, the chain reaction would produce more and more heat. This would damage the reactor and could be extremely dangerous. To prevent this and to control the amount of electricity being produced, the reactor has control rods containing the elements boron and cadmium.

Boron and cadmium absorb neutrons. Lowering the control rods into the core stops the chain reaction.

The coolant
The heat given out during fission is carried away by a cooling fluid or **coolant**. Different types of reactor use different coolants. Some use carbon dioxide gas, others use liquid sodium and some use water. The hot fluid is pumped to a boiler where it turns water into steam. This is used to turn the turbines of an electrical generator.

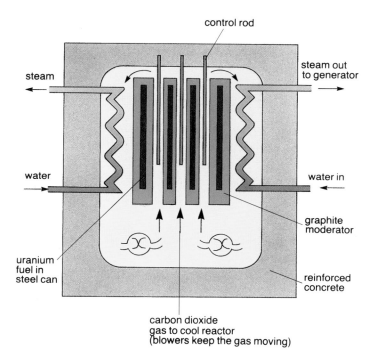

Design of the Advanced Gas-Cooled Reactor (AGR)

Questions

1 What is 'nuclear fission'?

2 Name the isotope of uranium which undergoes fission.

3 Name one element which slows down neutrons.

4 Name one element which absorbs neutrons.

5 Name one gas and one liquid which can be used to cool a reactor core.

6 Find out **a)** the fuel **b)** the moderator **c)** the coolant in each of these reactors: Magnox, Advanced Gas-Cooled, and Fast Reactor.

Dealing with waste

Radioactive materials are used in nuclear power stations, industry, medicine, and research. All these uses produce waste materials that are radioactive. They must be disposed of in ways which will not damage the environment or put human lives in danger.

Low level waste

This is waste which is contaminated with very small amounts of radioactive material. For example, in hospitals radioactive isotopes are used for treating certain diseases. The protective clothing worn by doctors and nurses, the glassware used to make solutions, and syringes used for injections are left slightly radioactive. These things are put into a fibre-board bin and then sealed inside a metal drum. The drums are then buried underground.

Intermediate level waste

Intermediate level waste would be dangerous to release into the environment because it has fairly high radiation levels. It includes the metal stripped from used reactor fuel rods, old parts from nuclear reactors, and equipment which has been used in highly radioactive places.

At present, intermediate level waste is sealed into steel drums and then encased in concrete. The concrete blocks are buried below the ground in solid concrete stores. Radiation levels in such stores will need to be monitored for about 300 years!

High level waste

This is nuclear fuel which has been used up inside a nuclear reactor. It is the most dangerous of all radioactive waste. It is highly radioactive and produces a lot of heat. It has to be stored under water to keep it cool. The level of radiation falls as the radioactive waste decays but it remains at a dangerous level for hundreds of years!

Most high level waste is stored at the Sellafield reprocessing plant where uranium and plutonium are recovered from used fuel rods. The waste is sealed inside stainless steel drums. The drums are then stored in a large tank of water. Eventually it will be possible to turn the waste into a glass-like solid. This will be kept for about fifty years before it is buried far below the earth in a **repository**.

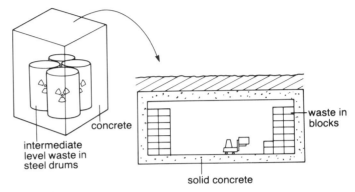

Storing intermediate level waste

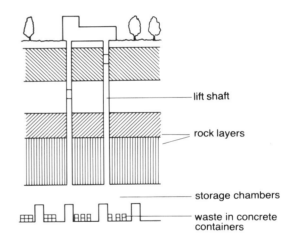

A repository built under the ground. Some may be built far below the sea bed.

Activities

1 Imagine that you own a house in a quiet country town. You hear that NIREX (the Nuclear Industry Radioactive Waste Executive) is planning to build a store for low-level nuclear waste near to your town. Write a letter to NIREX explaining why you are against the plans.

2 Now imagine that you are the managing director of NIREX. Write a letter to people living in the town explaining why you believe they will not be at risk.

Cleaning up

Moving waste about the country

After fuel rods have been inside a reactor for several years they are **reprocessed**. The old metal casing is cut off and then any unused fuel is separated from the waste. Any uranium goes to make new fuel rods. One of the decay products called plutonium is also recovered. This can be used in a different type of power station which uses a 'fast breeder reactor'.

Old fuel rods are highly radioactive. They are transported to the reprocessing plant at Sellafield in very special containers. These must not allow any spillage of radioactive waste even if they are involved in a serious accident.

The photograph shows a test carried out in 1984. A diesel train travelling at 100 mph was allowed to collide with a steel flask. The train was wrecked; the flask showed some damage on the outside but stayed in one piece.

Dealing with old reactors

The United Kingdom has nine nuclear reactors producing electricity at present. The oldest one started work in 1962 and will continue to produce electricity for several more years. Eventually the oldest reactors will need to be closed down and cleared away. This is called **decommissioning**. This will not be easy because the reactor will remain highly radioactive for many years. It may take up to 100 years to clear a site completely! It will also be very expensive; each power station will cost more than £200 million pounds to decommission. The diagrams show plans for removing an old power station.

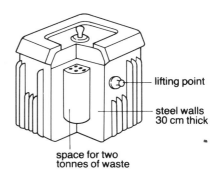

lifting point

steel walls
30 cm thick

space for two
tonnes of waste

This steel flask has a mass of about 50 tonnes. It will survive fire, water, and a train crash.

A spectacular test of a flask designed to carry nuclear waste.

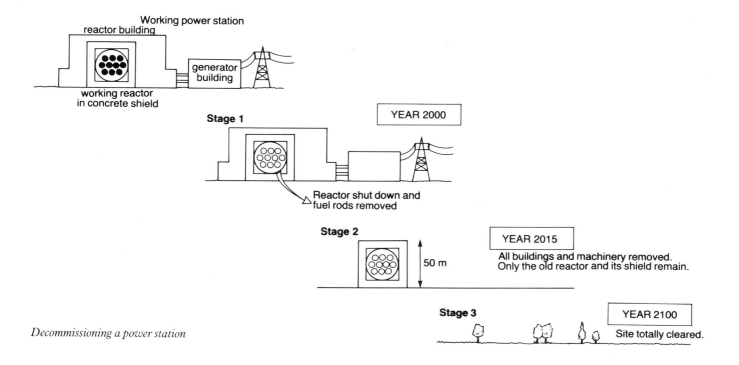

Decommissioning a power station

Working power station
reactor building

generator
building

working reactor
in concrete shield

Stage 1

YEAR 2000

Reactor shut down and
fuel rods removed

Stage 2

50 m

YEAR 2015

All buildings and machinery removed.
Only the old reactor and its shield remain.

Stage 3

YEAR 2100

Site totally cleared.

Nuclear power

Why use nuclear power stations?

Our world has become dependent on electricity. Each year we need more energy for use in our homes, schools, and factories. To meet this demand the Central Electricity Generating Board has built more power stations. Some of these burn oil or coal but for over 25 years nuclear power stations have been built in the United Kingdom.

Electricity from nuclear power is exactly the same as that made in a coal-fired power station. You can never tell when you are using electricity from uranium or from coal! Here are some of the reasons used to support the use of nuclear power:

1   A small piece of uranium produces very large amounts of energy. 1 tonne of uranium produces more electricity than 75 000 tonnes of coal!

2   Uranium is a very clean fuel. The reactor does not produce large amounts of waste gas which can pollute the atmosphere. (See Chapter 6.)

3   Nuclear power stations do not produce carbon dioxide gas as a waste product and so do not contribute to the greenhouse effect (see page 222).

4   By using uranium as a fuel, fossil fuel supplies are conserved for use elsewhere. This is very important in developing countries where coal and oil are needed.

Activities

1   Ask your friends and relatives whether they think we should build more nuclear power stations. Listen to their reasons. Do they use scientific arguments, for example, danger of accidents, efficiency of reactors, etc.? Do they use economic arguments, for example, cost of production of electricity, cost of clearing nuclear waste, etc.? Do they use political arguments, for example, the need for or dangers of nuclear weapons, etc.?

2   Look at information from the CEGB about the safety of nuclear reactors. Compare this with information from Friends of the Earth or other conservation groups.

3   Try to find out what went wrong at Chernobyl. Could it happen in the United Kingdom?

Is anything wrong with nuclear power?

Many people are against the development of nuclear power plants. There are several reasons for this:

1   Radioactive fuels and waste must be handled with great care. Any accidentally released into the environment could cause serious problems. It could cause sickness and death in humans, plants, and animals. The pollution could last for hundreds of years. Strict rules and special methods are used for transporting fuel and waste but many people are worried that there may be a serious accident.

2   Most of the reactors in the United Kingdom use uranium as a fuel to make electricity. One of the products inside old fuel rods is the element plutonium. This can be separated at the Sellafield reprocessing plant for use in fast-breeder reactors. However, plutonium is also used in nuclear weapons. Some people believe that using nuclear power to produce electricity will encourage the development of nuclear weapons.

3   There is a real fear that there will be an accident inside a nuclear reactor. The general safety record of nuclear reactors is very good but any accident would be very serious. This became a shocking reality in April 1986 when an explosion at a reactor in Chernobyl, Russia allowed a massive leak of radiation. Radioactive material was blown across Europe and farming land in Britain was polluted. Near the reactor 31 people were killed but, in the years to come, many people are likely to develop cancer and other diseases related to radiation. We will never know exactly how many people will die as a result of the Chernobyl disaster.

Chernobyl, Russia after the world's worst nuclear accident

Using isotopes

We have seen (page 103) the ways in which radioisotopes are used in medicine. These pages give examples of the uses of isotopes in industry and research.

Radiocarbon dating

All living things contain carbon. The common isotope of carbon is carbon-12 but there is a radioactive isotope, carbon-14. Plants and animals take in carbon dioxide from the air and so during their lives their cells contain a small amount of radioactive carbon-14. This decays but is replaced by more from the air. When the animal or plant dies, the carbon-14 trapped in the body decays but is not replaced.

The half-life of carbon-14 is 5700 years. If we know how much carbon-14 is left in the remains of a dead plant or animal we can calculate how long ago it died. This is called **radiocarbon dating**. It can be used to date the bones of dead animals or anything made from something which once lived. This includes cloth, paper and wood. The Dead Sea Scrolls are ancient religious and historical documents found near the Dead Sea, by accident, in 1947. They are made of leather and papyrus (primitive paper made from a plant) and were dated from their carbon-14 content. This proved they are about 1900 years old and therefore were written just after the death of Christ.

Investigating chemical reactions

In a chemical reaction all the isotopes of an element react in exactly the same way. They form the same compounds but each isotope can still be identified. To investigate what happens in a reaction scientists can 'label' an atom by using a different isotope.

Isotope labelling has been used to study photosynthesis in plants. The overall equation for photosynthesis can be written as

$$12H_2O + 6CO_2 \rightarrow C_6H_{12}O_6 + 6O_2 + 6H_2O$$
water   carbon dioxide   sugar   oxygen   water

You can see that oxygen gas is produced. Did this oxygen come from the water or from the carbon dioxide? To find out, plants are grown in normal carbon dioxide but are given water containing the oxygen-18 isotope instead of oxygen-16.

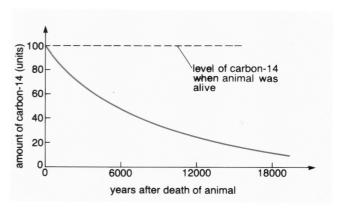

Graph showing how the amount of carbon-14 in an animal bone decays after the animal dies

Dead Sea scrolls

The oxygen gas given off by the plant also contains oxygen-18! This shows that the atoms in the oxygen gas come from the water molecules and not from the carbon dioxide.

$$12H_2O^\star + 6CO_2 \rightarrow C_6H_{12}O_6 + 6O_2^\star + 6H_2O$$
'labelled oxygen'   'labelled oxygen'

Questions

1 Archaeologists found some ancient gold jewellery. Explain why they could not use carbon-14 dating to find out how old it was.

2 A modern bone contains 100 units of carbon-14. A fossil bone contains just 25 units. How old is the fossil?

3 Give two reasons why carbon-14 is very useful for dating ancient things.

More uses

Using penetrating power

In factories making plastic sheeting radioactivity can be used to measure the thickness of the plastic as it is made. A beta source is placed above the plastic and a detector is placed directly below. Most beta particles can get through the plastic to the detector. If the plastic is too thick, fewer beta particles can get through. If the plastic is too thin then more beta particles reach the detector. The signal from the detector can sound a warning or can be used to adjust the machinery making the plastic.

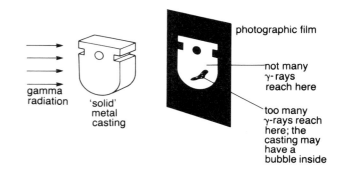
Monitoring material thickness in industry

Checking welds and castings

If a gamma source is placed on one side of a piece of metal, most, but not all, of the radiation gets through. If the metal is thinner the count rate on the other side is higher. This can be used to test welds joining two pieces of metal. It can also be used to find out whether metal castings are solid. If the casting has an air bubble inside more gamma radiation can get through.

Checking metal castings

Following liquids

Radioactive isotopes can be used to study the flow of liquids. For example, if an underground pipe is leaking some slightly radioactive liquid can be pumped into the pipe. A detector is then used above the ground. Where the liquid leaks out the detector gives a high count rate. This tells the engineers where to dig before fixing the pipe.

Radioactive isotopes can also show the transport of dissolved chemicals in plants. A liquid labelled with a radioisotope is used to 'water' the plant. The roots absorb the dissolved chemicals which can then be traced as they move through the plant.

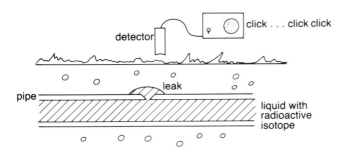
Finding a leak in a pipe

Questions

1 Why couldn't **a)** alpha particles or **b)** gamma radiation be used to monitor the thickness of plastic sheeting as it is made?

2 Beta particles passing through a sheet of plastic 0.1 mm thick give a detector reading of 50 counts per second. Suddenly the detector reading goes up to 70 counts per second. Suggest what may have happened.

3 Why is gamma radiation used to test metal castings?

4 Can you think of another way of checking whether a piece of metal has an air bubble inside? Would your method be as good as the gamma radiation method? Explain your answer.

5 An underground oil pipe running through the countryside starts to leak. Give two reasons why it is important for the leak to be found and mended quickly.

6 Explain why radioisotopes used for finding leaks in pipes should not have very long half-lives.

Nuclear fusion

Where does the energy come from in a nuclear reaction?

If we look at the equation for the splitting of a uranium-235 nucleus it is difficult to see where the energy has come from. It is as though we are getting it for nothing. It took the genius of Albert Einstein to solve this problem. He suggested that in a nuclear reaction a very small amount of mass is lost. It is this tiny bit of 'missing' mass which has turned into energy!

This was a very surprising idea. Until that time scientists believed that mass and energy were totally different things and that mass could not be destroyed.

Einstein went further by writing an equation which connects energy to mass: $E = mc^2$
(E = energy, m = mass, c = speed of light = 300 000 000 m/s)

In fact the mass lost when a single nucleus decays is very small; much less even than the mass of a proton, but the energy given out is quite large. For example 1 kg of radium nuclei decays to give 999.98 g of decay products. The 'missing' 0.02 g appears as 2 000 000 000 000 J of energy!

Albert Einstein (1879–1955)

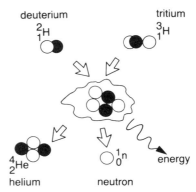
1 kg is a lot less than 60 tonnes!

1 kg of uranium produces as much energy as 60 tonnes of coal!

Activities

1 Try to find out about Albert Einstein.
   a) Where was he born and when?
   b) Was he a good student at school?
   c) What job did he do before he was recognized as a scientist?
   d) When was he awarded the Nobel Prize for physics?
   e) In 1939, Einstein wrote a letter to Franklin D. Roosevelt, President of the United States. What was it about?

Nuclear fusion

So far we have looked at what happens when a heavy nucleus splits up. However, we can also get energy when two light nuclei join together. This is what happens in our sun and other stars.

The diagram shows how two hydrogen nuclei join together to form a helium nucleus and a neutron. Once again a little bit of mass is 'lost'. This appears as energy.

Nuclear fusion

Trying to control nuclear fusion

Nuclear fusion would be a good way of generating electrical power on Earth because the fuels can be made from lithium and water, both of which are in plentiful supply and safe to use. Unfortunately it is very difficult to get the hydrogen nuclei to react. Huge numbers of hydrogen atoms must be heated to a temperature of about 100 million degrees centigrade; hotter than the Sun itself!

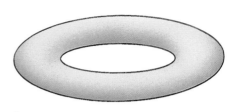

A torus

To get these high temperatures the hydrogen nuclei and their electrons are heated using electromagnetic fields. The magnetic field also keeps the hot, hydrogen 'plasma' away from the sides of the vessel where it is being heated. This is a very difficult thing to control but scientists have been experimenting with it for several years. One project is in Oxfordshire and is called **JET**. JET stands for Joint European Torus. Torus is the name given to the ring-shaped vessel where the plasma is heated.

The JET project has successfully produced a fusion reaction but it will be many years before we know whether a fusion reactor can be used to produce electrical energy but the idea is very attractive. The fuels would be cheap and readily available, the fuels would be safe and there would not be any radioactive waste.

The diagram shows the equipment being used in Oxfordshire to study nuclear fusion.

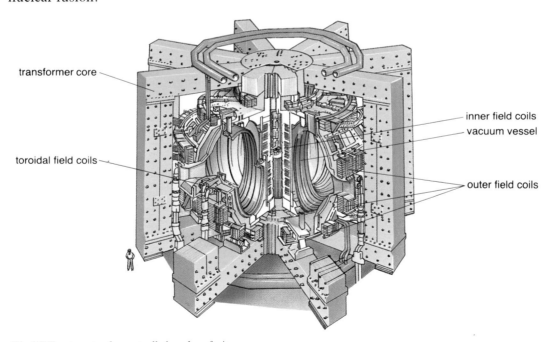

The JET apparatus for controlled nuclear fusion

Questions

1 Using the size of the man on the diagram to help you, estimate the total height of the JET apparatus.

2 How big roughly is the 'toroid' in which the hydrogen plasma is heated?

3 Suggest three reasons why nuclear fusion reactors would be better than the nuclear fission reactors we use now for generating electricity.

112

Questions

1 The diagram below represents an atom.

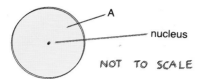

NOT TO SCALE

a) What type of particles can be found in the space labelled *A*?
b) Name *two* types of particle which can be found in the nucleus.
c) Suggest why the artist has written '*Not to scale*' under the diagram.

2 The table below gives the symbols and structures for some nuclei.

symbol	atomic number	nucleon number (mass number)	number of protons	number of neutrons
^1_1H	1	1	1	0
^4_2He	2	4	2	2
$^{13}_6C$	6	13	6	7
$^{16}_8O$	8	16	8	8

a) Is the atomic number written near the top or bottom of the symbol?
b) What other column is always equal to the 'atomic number' column?
c) What does the nucleon number (mass number) tell you?
d) Using the same headings make a table for the following nuclei:

$$^3_1H \quad ^{14}_6C \quad ^{235}_{92}U$$

3 A school radioactivity kit contains a cobalt-90 source which emits gamma radiation.

a) Draw the sign which should appear on the outside of the box of the radioactivity kit.
b) Suggest why it is important to have an *international* sign for radioactive materials.
c) The cobalt-90 is kept inside a box lined with lead. Explain why lead is used.
d) Write down two precautions which a teacher should take when using the cobalt-90 source.
e) A teacher is worried about the dose of radiation received when teaching the radioactivity topic. Suggest one way of 'measuring' the dose.

4 A student puts a sample of a radioactive mineral close to a detector. The count rate is very high. When a sheet of aluminium is placed between the sample and the detector, the count rate drops to a value close to the background count rate.

a) What is meant by 'background' radiation? Give two sources of background radiation.
b) How do you know that the mineral is not emitting gamma radiation?
c) Explain how you could find out whether the mineral is emitting alpha particles only, beta particles only, or alpha and beta particles.

5 Uranium-235 undergoes nuclear fission when it absorbs a neutron. This can start a chain reaction.

a) Use a diagram to explain what is meant by a 'chain reaction'.
b) In one type of nuclear bomb the 'fuel' is almost pure uranium-235. The fuel in a nuclear reactor is 97 per cent uranium-238 and 3 per cent uranium-235. Suggest why a nuclear reactor could not blow up like a nuclear bomb even if something went wrong.

6 The diagram shows how radioactive waste can be sealed inside a steel drum before dumping it into a deep part of the Atlantic Ocean.

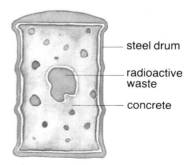

steel drum
radioactive waste
concrete

a) Suggest two reasons why conservation groups have tried to stop the dumping of nuclear waste at sea.
b) Describe how intermediate level waste can be stored in a deep repository.
c) When looking for a site to build a repository what things do you think should be considered?
d) Explain why it would be dangerous to burn radioactive waste.

6 The environment

What is the environment?
What happens to water in our environment?
What is in the soil?
How do we damage our environment?

'**Environment**' is a scientific word for 'surroundings'. You are probably reading this in your school environment or your home environment.

Your environment provides you with such things as air to breathe, water to drink and a suitable temperature in which to live. These are the **physical** or non-living parts of the environment.

Your life is also affected by other living things. These could be the people in your class, your family, your pets and even the bacteria in the air.

Living things, together with their physical environment, form an **ecosytem**. In an ecosystem many different processes or cycles take place. These help to keep the environment the same as years go by. We have to be careful that we do not upset the balance in our environment. The photographs show some activities which change the environment.

Building

Quarrying

Farming

Questions

1 Describe the environment where you are reading this book.
Give examples of the living and non-living parts of your environment.

2 Describe the environment you would choose for an ideal summer holiday.

3 The Council opens a large rubbish tip. How would it affect your environment if you lived nearby? How could it affect your environment if you were a seagull? (*Hint:* Seagulls eat scraps.)

Waste disposal

Some ecosystems

This aquarium contains tropical fish and plants. There will also be smaller living things, such as water snails and algae in the tank. The environment for this community of animals and plants is the water, the gravel at the bottom of the tank and the air at the top.

The aquarium is a small ecosystem.

An aquarium is not usually a '**closed** ecosytem'. The fish need extra food added by a person.

An aquarium; a small ecosystem

This pond is a **natural** ecosystem. Like the aquarium the plants and animals live in the same environment, the water, the soil and the air above.

In a well-balanced pond, the environment stays roughly the same. The plants do not grow too much and choke the pond nor do the animals eat all the plants or each other!

A pond; a natural ecosystem

This bottle garden is a small ecosystem. It is very good for growing plants like ferns which need moist conditions. The small plants are planted in soil inside the bottle. Some of the water in the soil evaporates but it cannot escape because the bottle has a narrow neck. This makes the air inside the bottle moist. The plants grow well because they have ideal 'weather' conditions in their environment.

Bottle gardens are not well-balanced ecosystems. The plants grow very well and soon fill up the bottle!

A bottle garden

This woodland is another natural ecosystem. Birds, insects and mammals live and feed in the wood. The trees provide food and shelter. As old trees die they will be replaced by new ones.

Once, nearly all of Britain was covered by woodland like this. As our ancestors cut down trees for firewood and to build houses, most of our woods disappeared. The woodland creatures died because the environment had changed too much.

Questions

1 a) What is a closed ecosystem?
   b) Why is a tank of tropical fish not a closed ecosytem?
   c) Suggest one other example of an ecosystem that is not closed.

2 Ponds, lakes and rivers are natural ecosystems. Explain what might happen if anglers took home all the fish they caught when fishing.

3 Suggest how a bottle garden could be modified to stop it becoming full up with plants.

4 The Earth is an ecosystem.
   a) Is it a closed ecosystem? Explain your answer.
   b) Give three ways that humans have harmed ecosystem Earth.
   c) Why must we take care to look after our 'planet environment'?

Natural woodland

Food chains: an introduction

Green plants make their own food during photosynthesis (see page 124) using energy from sunlight. Any living thing that makes its own food is called a **producer**. Green plants are therefore producers.

Animals, on the other hand, cannot make their own food. They must get their food by eating plants or by eating other animals. Animals are called **consumers** because they eat or consume other living things.

Animals that eat plants are called **herbivores**. Rabbits are herbivores. They eat grass, cereals, carrots, lettuce and many other plants.

Animals that eat other animals are called **carnivores**. A fox is a carnivore. It feeds on animals such as insects, mice, birds and rabbits.

Some animals feed on a diet that includes both plants and other animals. These are called **omnivores**. Badgers are omnivores. They feed on things like grass, fruit, slugs, worms and rabbits.

A badger; an omnivore

Each plant and animal is a link in a **food chain**. Energy, in the form of food, passes from producers to herbivores to carnivores.

Here is an example of a food chain.

A food chain with three links

Notice there are three links in this food chain. Not all food chains have three links. Some have more like this one.

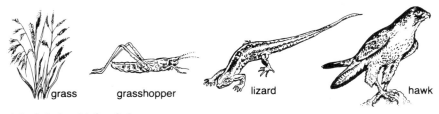

A food chain with four links

Questions

1 What is **a)** a producer **b)** a consumer? Give one example of each.

2 Are you a herbivore, a carnivore or an omnivore? Explain your answer.

3 Explain why all living things depend on sunlight for their food.

4 A student went to study a local pond. In one part of the pond she noticed tadpoles scraping at some pond weed. In another part she saw a water beetle holding a tadpole in its jaws.
  a) Construct a food chain for the pond.
  b) How many links are there in this chain?
  c) Suggest another food chain linking the following pond organisms:
  water fleas, perch, microscopic plants, stickleback.

5 Construct **a)** a long **b)** a short food chain ending with **you**.

Food webs

Single food chains do not give us a full picture of the feeding relationships between plants and animals. We all know that rabbits do not feed entirely on lettuce and foxes do not eat rabbits all the time, and you don't live on just fruit!

If we were to trace every food chain involving lettuce, rabbits and foxes we would end up with lots of interconnecting food chains. This is called a **food web**.

See how complicated it all can get with even a few living things in a food web!

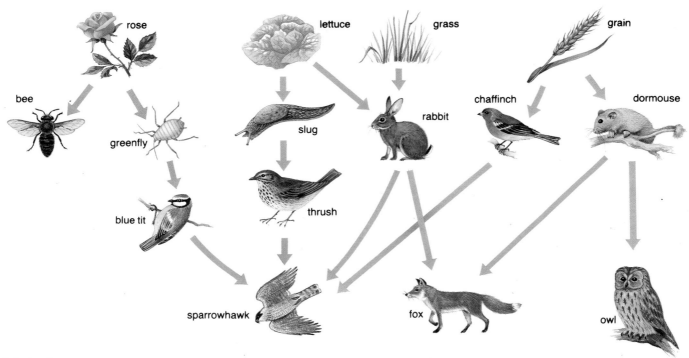

A food web

See how complicated it can get with even a few living things in a food web!

Questions

1 Look closely at the food web above then:
   a) Write down a food chain with   **i)** two links   **ii)** three links   **iii)** four links.
   b) Write down the names of the herbivores.
   c) Which carnivore has the most varied diet? Explain your answer.
   d) Suggest one way in which the sparrowhawk can   **i)** help   **ii)** hinder a gardener.

2 Rose growers often spray their roses with insecticide to kill greenfly.
   a) Explain why this can upset the food web.
   b) Suggest a way in which rose growers could control the numbers of greenfly without using any chemical sprays.

The water cycle

Living things need water. Humans use it for drinking and for a large number of other things. In fact a person in Britain uses, on average, 120 litres of water each day! The table shows how we get through so much.

The water in our environment is not used up. It is always on the move. This movement of water is called the **water cycle** and this keeps the water in the Earth's ecosystem in balance.

WC flushing and waste disposal	39 litres
Personal washing and bathing	37 litres
laundering	20 litres
dishwashing and cleaning	12 litres
gardening	4 litres
drinking and cooking	6 litres
car washing	2 litres

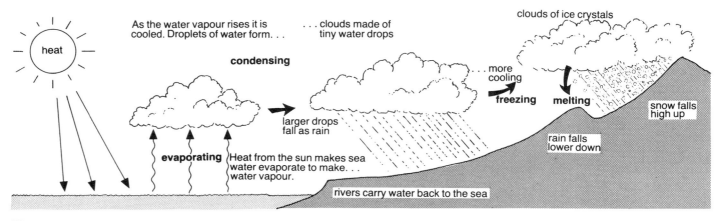

The water cycle

Where is the water in our environment?

. . . *2% is in the polar ice caps*

. . . *97% is in the oceans*

. . . *1% is in the lakes, rivers and the ground.*

Questions

1 Use the table on this page to answer the following questions:

a) What human activity uses up most water each day?

b) What activities do you think could be stopped at times of severe water shortage?

c) How much water on average does a family of four people use in **one year**?

2 Why does the level of the sea remain roughly the same?

3 What would happen to the amount of water trapped as **ice** if our planet became colder? Suggest how this would affect the water cycle.

Activities

Keep a record of the water you use in **one week**. You may need to use these estimates.

One cup = $\frac{1}{3}$ litre

One washing up bowl = 10 litres

One 'flush' of the toilet = 10 litres

One bath of water = 150 litres

One shower = 30 litres

Record your results in a table.

Work out the total amount of water you use in one week.

Plants need water too

All living things need water for the chemical reactions that take place inside them, to transport materials around the body and to carry away waste. In addition, plants use water to carry minerals from the soil into their roots.

How do plants take in water?

Water enters plants through the roots. The tips of roots are covered in tiny hairs. These root hairs are extensions of the outer cells of the root and grow out between the soil particles where water is held.

Water enters the roots by a process called **osmosis**. Osmosis is the diffusion of water from a weak to a strong solution through a selectively permeable membrane.

Root hairs take in water by osmosis because their cell contents are stronger than the surrounding soil and because their cell membranes are selectively permeable.

As root hair cells take in more and more water by osmosis their cell contents now become weaker than the cells next to them. So, water moves across into cells lying further inside the root.

If water continually enters a root system by osmosis it must go somewhere. Running through a plant is a miniature pipeline made of special cells called **xylem** and **phloem**. Xylem carries water and phloem carries food. It is into the xylem pipeline that water goes and as more water enters a pressure is built up. This root pressure helps move water through the plant, particularly in spring when plants begin to grow again.

Root pressure is only strong enough to push water a short distance up a stem. To get water the top of a plant requires the 'suction' effect of transpiration which you can read about on page 120.

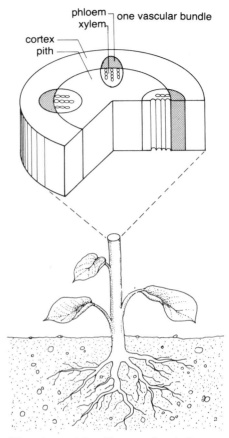

The xylem and the phloem are found all through the plant. They are held together in a vascular bundle

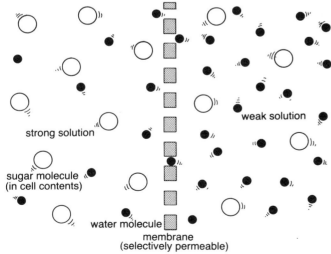

Since there are more water molecules in the weak solution, water will move through the tiny holes in the membrane into the strong solution. This continues until both sides are of equal strength.

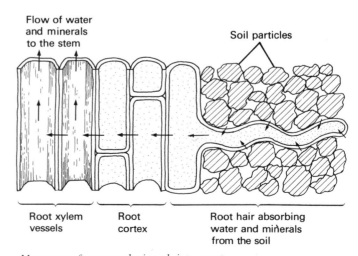

Movement of water and minerals into a root

Transpiration: water lost from plants

Transpiration is the process by which plants lose water vapour by evaporation into the atmosphere. The water passes through tiny holes called **stomata** which are found mainly on the lower side of leaves. Two sausage-shaped **guard cells** surround each stoma and control the size of the hole. If the pressure inside the guard cells is high the stoma will be open allowing gases to be exchanged with the air. However, if the internal pressure is low the stoma will close. This will prevent any movement of gases into or out of the leaves.

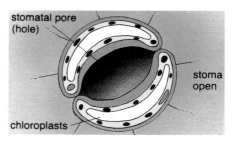

Stoma open

The rate of transpiration depends on a number of things:

- wind – on windy days water vapour will be blown away as it passes out of leaves.
- temperature – the warmer it is the more water will evaporate into water vapour.
- humidity – dry air can hold much more water than air that is wet.
- time of day – stomata usually close at night.

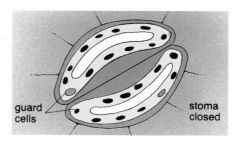

Stoma closed

The movement of water in the xylem from roots to leaves is called the **transpiration stream**. It appears that water is 'pulled' up the xylem tubes in a plant as water vapour is lost from the leaves during transpiration – just like a drink is pulled up a drinking straw when you suck on it. This transpiration pull is the main way in which water is transported in plants.

Investigating how external conditions affect transpiration

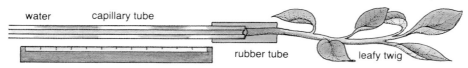

A potometer

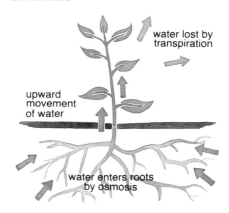

The transpiration stream

This apparatus is called a **potometer**. It can be used to measure the rate of transpiration in a leafy shoot. This is done by measuring how far the water moves along the capillary tube in a certain time, say one minute. By placing the potometer in different conditions we can see how the rate of transpiration is affected. Look at these results.

conditions	transpiration rate mm/min
warm place	15
cold place	5
still air	8
moving air	20
dry air	10
moist air	4

Questions

1 What is transpiration?

2 What are **a)** stomata **b)** guard cells? Explain how guard cells change the size of stomata.

3 Explain how **a)** temperature **b)** wind **c)** humidity affect the speed by which a plant loses water to the atmosphere.

4 Using the results of the potometer experiment, which conditions cause plants to lose water **a)** fastest **b)** slowest?

5 Explain why a potted plant placed on a sunny window sill will soon start to **wilt**.

6 A student says 'Plants lose water in the same way as clothes dry on a washing line'. Is the student correct? Explain your answer.

Too little water

Plants and animals need water from their environment but some environments provide very little water. In the Sahara desert the average daytime temperature is about 35°C and it only rains about once every three years!

Not all places are as bad as the Sahara but even so some plants and animals have adapted to survive in very dry conditions.

Survival is only possible if animals and plants can prevent themselves drying up.

Some desert animals avoid losing too much water from their bodies by staying out of the sun. Gerbils are a good example. They live in burrows during the day and come out at night to feed when it is much cooler. They get all their water from the food they eat because there is very little water to drink.

Camels are very well adapted for desert life. When they get the chance to drink at an oasis or at a well they can take in lots of water very quickly; over 100 litres in 10 minutes! They can then go for many days without drinking.

Camels' humps store food as fat. This can be used when there is no food to eat. Camels are large animals and this helps them survive. During the cold desert nights they let their body temperature drop by several degrees. During the day they heat up slowly. When the evening comes the camel's temperature is just a few degrees above normal.

Like most desert animals, camels do not waste water by producing lots of urine. Their droppings (faeces) are dry when they have been without water for some time.

The best known desert plants are cacti. They are very well adapted to dry environments. The most obvious feature is that their leaves are reduced to **spines**. The spines have a small surface area so that less water is lost. Cacti also have stomata that are closed during the hot day and open at night when water loss will be less. (The trees in Britain have their stomata open during the day and closed at night.) Cacti also have shallow root systems which spread out a long distance from the base of the stem. The roots can then take in any dew which forms on the ground. This water is stored in the green stem of the plant which becomes fleshy and even takes over the role of leaves in photosynthesis.

A cactus; well adapted for the desert

A camel in a desert environment

Activities

Get two or three plants of roughly the same size which are growing in pots. One should be a plant with leaves. A busy lizzie (*Impatiens*) would be ideal. Another should be a cactus or a succulent. The jade plant (*Crassula argentea*) is a good example. Water the plants thoroughly and put them on a window ledge. Look at them every day but do not water them. Make a note of any changes that you see.

Answer these questions:
1 a) Which plant wilts first?
   b) Why do think this plant lost water fastest?
2 a) Which plant needed least water?
   b) How is this plant well adapted for a dry environment?

The nitrogen cycle

The element nitrogen is essential for life because it helps to form protein, the material of which all living things are made.

The air in our ecosystem is nearly 80 per cent nitrogen gas. Unfortunately animals and most plants cannot use nitrogen gas from the atmosphere directly. Plants must have their nitrogen combined with oxygen in the form of nitrates which they absorb from the soil. Animals get theirs by eating plants. In this way nitrogen passes along food chains. Eventually nitrogen passes back into the air.

The circulation of nitrogen compounds in the environment is called the **nitrogen cycle**.

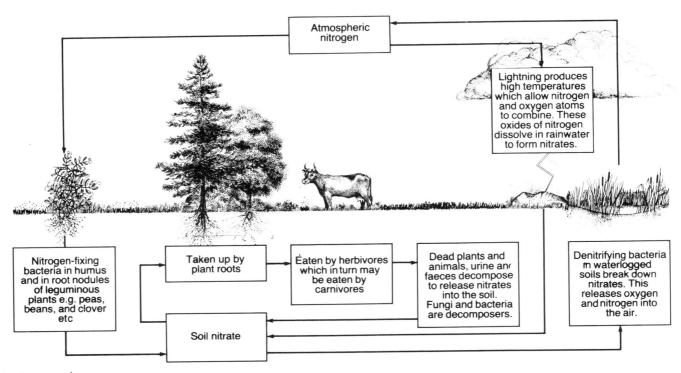

The nitrogen cycle

Questions

1 Explain how thunderstorms can make soil more fertile.

2 Often gardeners put dead plants on a compost heap. When the plants have rotted, the compost is dug into the soil. Why does this help next year's plants to grow well?

3 Peas are *legumes*. The roots of pea plants have nodules on them.
  a) What are the nodules for?
  b) Sometimes farmers grow pea plants and then plough them into the soil. Why does this help next year's crop of corn to grow well?

4 Old leather shoes left in damp, warm conditions go mouldy. The mould is a fungus. Explain how the shoes have become part of the nitrogen cycle.

Mould (a fungus) growing on leather

Minerals from the soil

Plants need nitrogen and other elements for normal, healthy growth. Mineral salts from the soil enter the roots of plants in the form of ions in solution. Some ions may enter by simple diffusion but most get into the cells of the root by a process called **active transport**. Active transport uses energy to move ions into the root cells. It means that plants can take ions into their cells when there are more ions in the cells than in the soil (see page 119).

If plants are continuously removing ions from the soil, where do replacements come from? Clearly vegetation would die if the mineral salts were not replaced.

Some minerals come from rocks. Rain-water is slightly acidic because it has some carbon dioxide from the air dissolved in it. This slowly dissolves the rocks and so salts from the rocks are washed into the soil. Other mineral salts come from the faeces of animals and the decomposition of dead plants and animals.

Fertilizer to improve plant growth

Mineral elements needed by plants

element	ions in the soil	what it is used for
nitrogen	(NO_3^- nitrate ions)	a vital part of protein for growth
calcium	(Ca^{2+} ions)	important for making cell walls
iron	(Fe^{2+} and Fe^{3+} ions)	for making chlorophyll for photosynthesis
magnesium	(Mg^{2+} ions)	for making chlorophyll for photosynthesis
phosphorus	(PO_4^{3-} phosphate ions)	a part of genes, chromosomes and cell membranes
potassium	(K^+ ions)	involved in enzyme action
sulphur	(SO_4^{2-} sulphate ions)	used in important proteins

There are other elements that plants require but only in very small quantities. These trace elements include manganese (Mn^{2+}), zinc (Zn^{2+}) and copper (Cu^{2+}).

The plant on the left is lacking in nitrogen

Questions

1 Where do plants get their mineral salts from?

2 How does active transport help a plant take in mineral ions from the soil?

3 How does oxygen help a plant obtain mineral salts?

4 A gardener finds that the cabbages in her garden have small leaves. Explain why a fertilizer rich in nitrogen would help.

5 In some soils the leaves of rhododendron plants turn yellow. If the soil near the plant is watered with a solution of a chemical called *sequestrene* the leaves look green again after a few days.
    a) What chemical, needed for photosynthesis, gives leaves their green colour?
    b) Suggest one ion which *sequestrene* is likely to contain.

Photosynthesis: leaves for the job

Trees, bushes, grass and flowering plants make our environment look attractive but they have a much more important role. They are an essential part of our balanced ecosystem. Plants are part of our food chains so without them we would die!

Plants make their food from non-living things by a process called **photosynthesis**. The raw materials used are carbon dioxide from the air and water from the soil. These are combined to form carbohydrates like sugars and starch.

Photosynthesis uses energy from the sun. The chemical **chlorophyll** 'traps' this energy.

Oxygen is a product of photosynthesis. This is needed by other living things in our ecosystem.

Leaves are the 'food factories' of plants. They are where photosynthesis takes place. They are well adapted for their job. Their broad, flat, thin shape provides a large surface area, ideal for the absorption of carbon dioxide and sunlight.

Green plants need sunlight.

Questions

1 What is photosynthesis?

2 How does the shape of leaves help them do their job?

3 What is the purpose of the layer of wax (cuticle)?

4 Why do you think the upper skin of a leaf is transparent?

5 Why do you think there are more chloroplasts in the upper part of the leaf than in any other cells?

6 There is also a film of water around leaf cells. What do you think it is for?

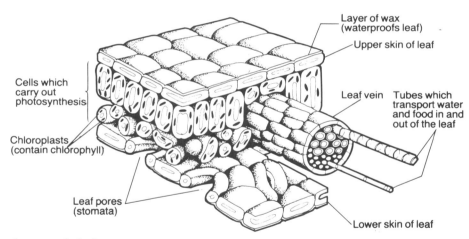

Structure of a leaf

Activities

Collect one leaf from six different trees or bushes.

Look carefully at each leaf then record your results in a table like the one shown below.

Can you find any patterns in your results?

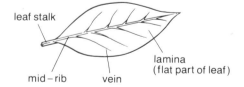

Name of plant	Top surface of leaf	Bottom surface of leaf
Privet	Dark green, shiny, smooth to touch.	Lighter green. Not so shiny or smooth. Petiole stands out on this side.

124

Light for photosynthesis

Rainbows show that light is made up of seven colours:

Red, Orange, Yellow, Green, Blue, Indigo and Violet.

These are the colours of the spectrum.

You can split light up in the laboratory like this:

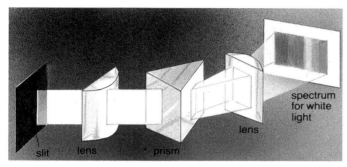

If you place some chlorophyll extract in between the light source and the first lens, the spectrum looks like this:

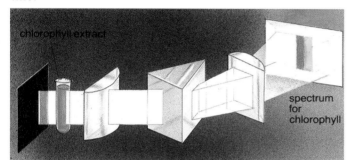

Some of the colours have 'disappeared'. This is because they have been *absorbed* by the chlorophyll. This shows that plants use only some of the colours of the spectrum when making glucose. The red and blue ends of the spectrum are absorbed but the middle part is not. The green light is reflected back. This is why chlorophyll and leaves are green!

Limiting factors

The intensity of light is one of the factors that has an effect upon the rate of photosynthesis.

As the light intensity rises the rate of photosynthesis increases only to a level permitted by the concentration of carbon dioxide in the surrounding atmosphere. In this example carbon dioxide is the limiting factor. If the carbon dioxide concentration rises but the light intensity is kept low then light becomes the limiting factor.

Variegated leaves; where is the chlorophyll?

Questions

1   What are the colours of the spectrum?

2   Which colours of the spectrum do plants use when making food?

3   Why are leaves green?

4   Suggest another limiting factor for photosynthesis. (*Hint*: Remember that photosynthesis is a series of chemical reactions controlled by enzymes.)

Activities

Collect the leaves of plants which are not entirely green. You should be able to find variegated leaves. These have green parts and white or yellow parts. You should also be able to find plants with red-looking leaves (e.g. Copper beech tree) and silver-grey leaves. Test the leaves for starch. Do they contain chlorophyll?

Photosynthesis: what happens inside cells

Photosynthesis is not one, simple chemical reaction, it is a series of complicated reactions. Each reaction is controlled by special chemicals called **enzymes**. (You can read about enzymes in Chapter 9.)

We can represent what happens during photosynthesis by one equation:

$$6CO_2 + 6H_2O \xrightarrow{\text{Sunlight absorbed by chlorophyll}} C_6H_{12}O_6 + 6O_2$$

carbon dioxide   water   glucose   oxygen

Usually the glucose made in this way is converted to **starch**. This is then stored for a short time in the leaf until it is needed.

Testing a leaf for starch

The starch test is very important because we can test the leaves of plants which have been kept under different conditions. A leaf which has been growing in the dark will not contain starch. Neither will a leaf that has had its supply of carbon dioxide cut off. Variegated leaves only contain starch in the areas where there is chlorophyll.

Where does the oxygen come from?

If you look at the simple equation for photosynthesis you will see that the oxygen given out by the plant could have come from the carbon dioxide (CO_2) or the water (H_2O). To investigate this, scientists water the plants with special water which has been 'labelled' with atoms of oxygen that are slightly heavier than normal oxygen (see page 109). (In fact they use the oxygen-18 (^{18}O) isotope in place of the normal oxygen-16 (^{16}O).)

When the oxygen gas given off is tested it is also 'labelled' with oxygen-18. This proves that the oxygen comes from the water and not the carbon dioxide.

Testing a leaf for starch

1 *Put a leaf into boiling water for about two minutes to soften it.*

alcohol

The flame must be put out. Alcohol burns!

2 *Heat it in alcohol to take away the green colour.*

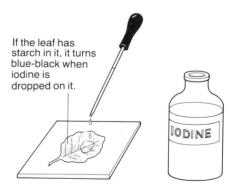

If the leaf has starch in it, it turns blue-black when iodine is dropped on it.

IODINE

3 *Soften it in boiling water, then add iodine to it.*

Questions

1 *Photo* means 'light' and *synthesis* means 'building up'.

  a) What chemical do green plants have to 'trap' the energy in sunlight?

  b) What large molecule is 'built up' during photosynthesis?

  c) What two simple chemicals are needed for photosynthesis?

2 **a)** Why do leaves change glucose to starch?

  b) How would you test a leaf for starch?

  c) How would you show that a plant needs sunlight for photosynthesis?

3 A scientist tests the air and finds that it contains 20.5 per cent oxygen and 0.05 per cent carbon dioxide. Would you expect the air inside a greenhouse full of plants to have the same amount of these gases? Explain your answer.

The carbon cycle

All living things contain the element carbon. In our ecosystem, carbon circulates from the environment into living things and then back again. This is known as the **carbon cycle**.

At the same time as carbon is circulating, oxygen is also passing in and out of living things. The carbon cycle and the **oxygen cycle** are connected.

Photosynthesis and respiration

We have seen that plants photosynthesize. This can be written as a word equation.

$$\text{Carbon dioxide} + \text{Water} + \text{Energy} \rightarrow \text{Food (sugar)} + \text{Oxygen}$$

Respiration is how animals and plants release energy from their food. (You can read more about this in Chapter 9.)

The simple word equation for respiration can be written

$$\text{Food (sugar)} + \text{Oxygen} \rightarrow \text{Carbon dioxide} + \text{Water} + \text{Energy}$$

Notice that the respiration equation is the opposite of the photosynthesis equation. In a balanced ecosystem, photosynthesis and respiration will keep the oxygen and carbon dioxide levels in the atmosphere steady.

Unfortunately it is not that simple. For example, large quantities of carbon are trapped as the fossil fuels coal, oil and natural gas. (See Chapter 4.)

When we burn these fuels carbon dioxide is given off. At the same time lots of the world's jungles and forests are being cut down. This means that there are less plants to photosynthesize and produce oxygen. These two activities may be upsetting the balance of the gases in our environment and so scientists must continually monitor what is happening in our atmosphere.

Every time you light a gas cooker you are releasing carbon atoms which were locked up millions of years ago!

Questions

1 A student says 'Respiration is the opposite of photosynthesis'.
a) Write the word equations for respiration and photosynthesis. Use these to show what the student means.
b) Give one reason why respiration is not *exactly* the opposite of photosynthesis.

2 A large aquarium contains fish and green plants.
a) What gas is given out when the fish respire?
b) What gas is taken in when the green plants *photosynthesize*?
c) What gas is given out when the green plants *photosynthesize*?
d) What gas is taken in when the fish *respire*?
e) The gases dissolved in the water are perfectly balanced during the day. What happens at night?

3 A large forest is cut down. The branches are then burnt. Give two reasons why this may increase the amount of carbon dioxide in the atmosphere.

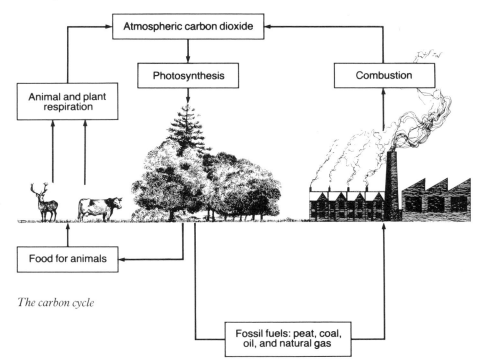
The carbon cycle

What is soil?

We see soil about us all the time. We see it in gardens, in fields and where holes are being dug in the roads. It is so familiar that we forget that the soil is an essential part of our environment. Without 'good' soil farmers cannot grow crops and millions die. It is very important to know exactly what soil is and how to look after it.

Soil is a mixture. It begins life as rock. The weathering action of wind, water and ice slowly breaks down the rocks into small particles (see page 63).

The stages of weathering can be seen in a **soil profile**. Try digging into the ground with a spade – the particles should get larger as you get deeper. If you could dig deep enough you would come to solid rock.

Soils are named according to the size of particles in them. **Clay soil** is made up of very small particles, **sandy soils** contain larger particles.

Rock particles make up the mineral or inorganic 'framework' of the soil. Mixed in with this is **humus**. This is organic matter made up of the remains of dead animals and plants. Humus 'binds' the rock particles together forming soil crumbs and prevents them from being blown away too easily by the wind.

Most artificial fertilizers do not have this 'glueing' ability.

Many of the essential mineral salts needed by plants come from humus.

Water covers the rock particles in a thin film. It is from this water film that plant roots take their water supply. When a soil is saturated with water it is said to be at **field capacity**. Clays soils have a high field capacity. They hold water and so they are hard to dig when wet.

The spaces between the rock particles in the soil contain **air**. Any organism living in the soil requires oxygen at some time and this is provided by the soil air. The air spaces are smaller in clay soil.

Farmers and gardeners prefer soils that are easy to plough or dig and are fertile. From what you have read you should realize that the 'ideal' soil will contain a good balance of clay and sand particles and have a plentiful supply of humus. Such a soil is called **loam**.

Soil profile

Activities

Looking at soil particles

1 Fill a glass container, such as a jam jar or large test tube, about a third full of soil.

2 Top up with clean water and secure the lid firmly.

3 Shake the mixture thoroughly for about a minute.

4 Leave the container for a few days in a safe place so that it will not be disturbed.

5 Look carefully at the way the soil particles have settled. They may look like this

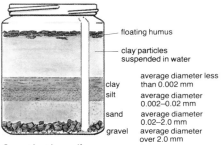

Investigating soil

How many different particle layers can you see in your container? Try repeating the activity with soil from different areas or with various types of compost.

Questions

1 Suggest why gardeners call sandy soils 'light' and clay soils 'heavy'.

2 A gardener tests the soil in her garden and finds that it contains very little humus.
   a) What is humus?
   b) What should she dig into the soil as *humus*?

3 Why is air an important part of the soil?

4 A gardener wants to grow plants in tubs. His garden contains clay soil. Suggest how he could make a good *loam* for the tubs.

5 Most 'modern' farmers use artificial fertilizers. There are however some serious disadvantages to this practice, can you suggest what these are?

The soil environment

Looking at organisms in soil

Fungi and bacteria are **decomposers**. They produce humus and break it down further to recycle important minerals and improve the soil fertility.

Soil fertility is also improved by earthworms. They add to humus by pulling leaves down into their burrows. They also let air into the soil as they tunnel deep into the ground. The familiar casts deposited on the surface of the soil by earthworms ensure the movement of 'fresh' soil to the surface like a small natural plough! It is estimated that there are about 50 000 earthworms in an acre of fertile soil.

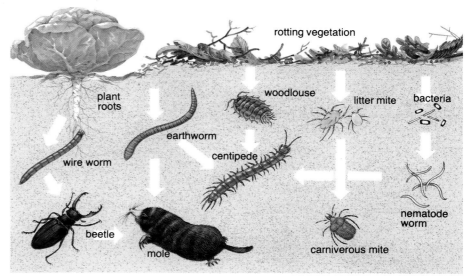

An underground food web

Other burrowing animals such as millipedes, a variety of insects, moles, rabbits and badgers all turn the soil over. They also add to the soil fertility with their droppings and, eventually, their dead bodies.

The soil environment also contains organisms that we regard as pests. These feed on other plants and soil animals. Wireworms, nematode worms, mites and parasitic fungi are examples.

Bacteria and fungi in the soil are a vital part of the nitrogen cycle referred to earlier in the chapter.

Activities

1 Shake a little soil with some boiled water in a test tube.
2 Put a few drops of this soil water onto a dish containing sterile nutrient agar jelly.
3 Seal the lid with sticky tape and keep the dish in a warm place.
4 Look at your dish after a few days.

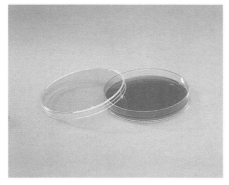
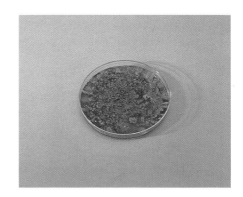

Growing microbes in the laboratory
(Colonies of bacteria appear as smooth blobs whereas fungi are 'furry'.)

Natural resources

This chapter has shown how carbon, water and nitrogen pass through the Earth's ecosystem over and over again. These cycles maintain a balance in levels of the resources.

Energy, however, does not circulate in ecosystems. It enters from outside, flows through and eventually leaves, usually as heat.

The source of energy for our ecosystem is the Sun. Energy is 'trapped' by green plants during photosynthesis. The energy-rich products of this process are available for food or fuel. When the food is digested or the fuel is burnt, heat is produced which can be lost from our ecosystem to outer space.

With the dramatic rise in the human population in recent years the demand for natural resources has increased enormously. As a result we are in danger of upsetting the delicate balance of cycles and using up valuable sources of energy.

Some resources, like water . . .

. . . soil . . .

. . . and timber . . .

. . . are renewable, but they must be carefully managed. If we are too greedy and use them up too quickly we will create shortages. Demand should never exceed supply!

Other resources, like fossil fuels . . .

. . . uranium for nuclear energy . . .

. . . and species of animals and plants . . .

. . . are non-renewable – once used, they are gone forever.

Questions

1 List some renewable resources.

2 List some non-renewable resources.

3 What is the difference between a renewable and a non-renewable resource?

4 Suggest why we should refer to soil as being a renewable resource.

5 In Britain thousands of trees are cut down every day to make paper for newspapers and magazines.

Suggest **three** ways to make sure that we do not destroy our timber supply.

Pollution

As we use more of our natural resources we produce more waste. The build-up of waste leads to pollution of our environment. Anything which damages our environment is called a **pollutant**. Pollutants can cause illness and pollutants can kill!

Our local environment can be polluted with . . .

traffic noise and exhaust gases . . .

. . . litter . . .

. . . and cigarette smoke.

The search for more fossil fuels and minerals results in the pollution of . . .

water by oil . . .

. . . and pollution of land by slag heaps.

By-products of industrial processes include sulphur dioxide and compounds of lead, mercury and cyanide.

Sulphur dioxide dissolves in rain forming an acid. This 'acid rain' falls on land and in lakes far away from where it is first released, destroying trees and other life forms. (See Chapter 4.)

Lead, mercury and cyanide compounds are poisonous even in small quantities. They accumulate in the bodies of animals, especially fish, mainly because they are dumped into rivers that flow to the sea. People eating contaminated fish will suffer serious illness and probably die.

The effect of acid rain

The waste from nuclear power stations can give off radiation for thousands of years.

Questions

1 What is a pollutant?

2 List some pollutants of your school environment. Suggest how these could be controlled.

3 List some pollutants of your town environment. Suggest how these could be controlled.

4 Some farmers use pesticides to kill unwanted animals, and herbicides to kill unwanted plants. Why do you think they do this? What dangers could result from the use of such chemical poisons?

Conservation

This chapter has shown how humans are part of an ecosystem which is delicately balanced. Unfortunately our activities upset this balance and cause serious problems.

At present we are demanding more and more from our natural environment. Eventually we will exhaust our resources. Some species of animal are now extinct due to humans and our gas and oil reserves will run out in less than 100 years.

We have a choice to make; either we carry on as we are or we try to **conserve** what we have left.

Conservation is a familiar word. It means taking care of our world so that it stays a fit place to live in. Already efforts are being made to combat the effects that we are having on our environment.

To conserve fossil fuels alternative energy sources have been developed. See Chapter (4).

Valuable materials can be recycled. For generations we have thrown away valuable materials such as metals, glass and paper. With a little effort these can be collected and reused. All it takes is the time to separate out such items from your everyday family rubbish and then take it to the appropriate place. Many towns and cities have civic amenities tips where rubbish can be separated for recycling. Bottle banks should be a familiar sight to us all.

Rare and endangered species of animals and plants stand a greater chance of survival with the establishment of nature reserves and National Parks. Local conservation groups are becoming more and more active, helping manage parts of our environment that might otherwise disappear for ever.

Already many laws exist to protect the environment but we must not just rely on others to do our thinking for us. If we want a pleasant place in which to live we must all help to maintain what we have.

Questions

1   What is conservation?

2   Give some examples of resources which can be recycled.

3   How can you help to conserve our environment?

4   If you were able to pass one law about the conservation of the environment, what would it be?

Activities

1   The emblem of the World Wide Fund for Nature (WWF) is a giant panda. Find out
   a) where the giant panda lives in the wild,
   b) how conservationists are trying to save the giant panda from extinction.

2   Elephants and black rhinos are endangered species in some parts of the world. Explain how humans are responsible for this. Find out how governments are trying to protect the remaining elephants and rhinos.

The dodo; killed off by Man

A bottle bank

Nature reserves protect plants and animals.

Emblem of the World Wide Fund for Nature

Questions

1 In South America many people are very poor. They cut down the trees of the jungle to sell. In their place they plant a crop, such as coffee, which they can sell for money to richer countries. With the money they buy food for their families.

   a) Explain why the South Americans need to cut down the jungle.
   b) Give *three* reasons why cutting down forests and jungles damages the world environment.
   c) Suggest *two* things which might help to solve the problem.

2 Rich countries like the United States and Great Britain have been called 'the throw-away society' because people throw away things which could be reused or recycle.

   a) List *three* things which we throw away which could be recycled.
   b) Explain why recycling waste paper helps to conserve our environment.

3 A gardener noticed that the leaves of his favourite climbing plant were much bigger on the shady side of his garden wall than on the sunny side.

   a) Suggest a reason for this.
   b) How would you test your theory?

4 'FIZZO' lemonade is sold in plastic bottles. 'LEMO' lemonade is sold in glass bottles.

glass   plastic

make	FIZZO	LEMO
price	28p	38p
money back on bottle	–	10p
weight of bottle (full)	1100 g	1500 g

a) Using information from the table suggest *two* reasons why a shopper might choose to buy 'FIZZO' lemonade rather than 'LEMO'
b) A person who wants to conserve the world's resources may choose to buy 'LEMO'. Suggest why.

5 Your family allotment has not produced very good crops for a few years, in fact only peas and beans seem to show any real growth.

If you had a supply of chemicals enabling you to make up a variety of solutions containing known mineral salts, devise an experiment to test whether the allotment soil is deficient in one or more mineral salts.

6 Space travel to another solar system would take many years. One idea is to build a space 'bubble' which would contain a sealed ecosystem where astronauts could live happily on a long journey. The diagram below shows one design for such a 'bubble'.

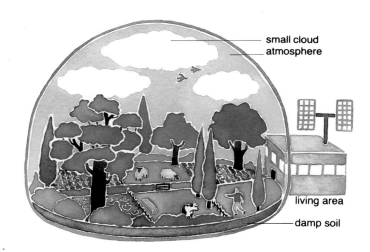

a) Suggest how **water** might circulate in this ecosystem.
b) i) Why do the animals in this ecosystem need oxygen?
    ii) Why do the plants need carbon dioxide?
    iii) How can the plants and animals contribute to a balanced atmosphere?
c) The space bubble travels so far from our sun that it becomes very dark inside.
    i) Explain what happens to the plants.
    ii) Explain what happens to the animals.

7 Electrical circuits

What happens in electrical circuits?
How do we use electrical circuits?
What is 'mains' electricity?

In our modern society we depend on electrical devices to do work for us. The photographs show some of the electrical appliances which we use in our homes. Each one converts electrical energy into another form.

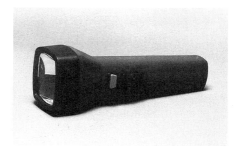
Converts electrical energy into light

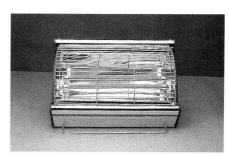
Converts electrical energy into heat

Converts electrical energy into kinetic energy and heat

Converts electrical energy into sound

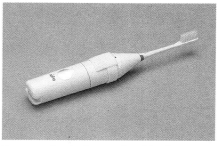
Converts electrical energy into kinetic energy

Converts electrical energy into light (and heat)

Notice that each appliance needs to be connected to a power supply. Some contain batteries. Others must be plugged into a socket connected to the mains supply.

WARNING

Remember that it is never safe to experiment with electrical circuits connected to the mains supply. Electric shocks can kill!

At home you can experiment safely with the type of 'battery' used in torches and bicycle lamps.

In the laboratory you may use a low voltage power supply or transformer which plugs into the mains. Always follow the teacher's instructions.

Activities

1 List all the electrical appliances in your home. For each one, state whether it uses batteries or is mains operated. Also write down the type, or types, of energy 'produced' by the appliance.

A table like the one below will help you to set out your answer.

appliance	mains or battery?	converts electrical energy to . . .
table lamp	mains	light (and heat)

2 What would life be like if we did not understand how to use electricity? Write some notes showing what you think the main changes would be. In particular think of home, transport, and communications.

Currents and circuits

All materials contain atoms. Atoms in turn contain small, electrically charged particles called electrons and ions. The diagrams show one way of thinking about these particles.

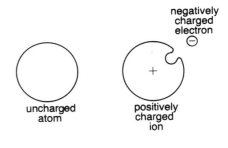

An electrical circuit is a pathway which charged particles can move along. For example, the diagram opposite shows a simple circuit for lighting a bulb. The circuit contains a battery. We think of the battery as 'pushing' the charge around the circuit. It is connected to the bulb by wires made of metal. The metal is a **conductor**; conductors let charge pass along them. Look at the circuit and you will see that there is a complete path from the positive end of the battery to the bulb, through the bulb and back to the negative end of the battery.

An electrical circuit must have a complete conducting pathway. If there are any gaps, the charge cannot flow.

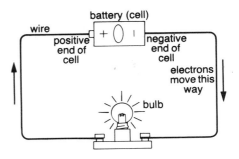

Switches

Switches are used to control electrical circuits.

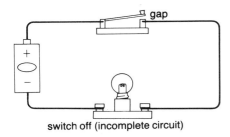

The bulb does not light up.

There is a gap in the circuit which will not conduct electricity. (Air is an **insulator**.)

We say that the switch is **open** *or* **off**.

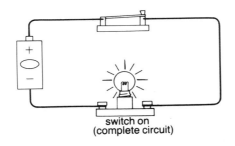

The bulb does light up.

The metal switch conducts electricity and so the circuit is complete.

We say that the switch is **closed** *or* **on**.

Questions

1 In the circuits opposite the wires are made of *copper* and are covered with *plastic*. The filament inside the bulb is made of *tungsten*. The switch has *brass* contacts and a *wooden* base.
For each of the five materials listed, state whether it is a conductor or an insulator.

2 Electric wires must be insulated. Some people insulate their houses. What is the difference between an *electrical insulator* and a *thermal insulator*?

Circuit diagrams

It is not convenient to draw pictures of electrical components all the time. Also, if you are a bad artist, other people may not understand your diagrams! Scientists and electrical engineers use **symbols** to represent the parts of an electrical circuit. The diagram opposite uses some of the standard symbols.

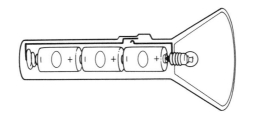

This drawing shows the inside of a torch.

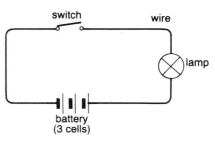

This is the circuit diagram for the torch.

Measuring currents

In a circuit containing a light bulb, we can judge the size of the current by looking at the brightness of the bulb. A large current makes the bulb glow brightly. A small current makes the bulb glow dimly or not at all. This method is not accurate enough for scientific work so a meter, called an **ammeter**, is used to measure the current.

The size of the current is measured in units called **amperes**. We nearly always shorten this to **amps** (symbol A).

When an ammeter is used it is connected **in** the circuit. It measures the current flowing through it.

An ammeter

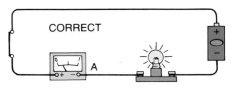

Here the meter is connected in the circuit. It measures the current in the lamp.

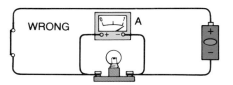

Here the meter is not connected in the main circuit. It does not measure the current in the lamp. The meter may be damaged.

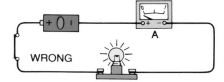

Here the meter is connected the wrong way around. The pointer moves the wrong way. The meter may be damaged.

What is an ampere?

The size of a current shows us how much charge flows through the circuit in one second. The unit of charge is the **coulomb** (symbol C).

A current of 1 A means that 1 C of charge flows in each second. (In fact each electron carries only a very small charge. About six million million million electrons are needed to give a charge of 1 coulomb!)

Sample question

What is the current in the circuit opposite?
Answer 2 A (Read the meter!)

How much charge flows through the lamp in one second?
Answer 2 C (1 A means 1 C in each second so 2 A means 2 C in each second.)

How much charge flows through the lamp in 10 s?
Answer 20 C (2 A means 2 C in each second so in 10 s we have 2×10 C.)

Hint: If you multiply the current (I) by the time (t) your answer is the charge (Q). In mathematical language ... $Q = I \times t$.

Questions

1 The diagram (right) shows an ammeter scale. What is the current when the pointer is at
a) A   **b)** B   **c)** C?

2 When the pointer is at B, how much charge flows through the ammeter in **a)** 1 s   **b)** 10 s   **c)** 60 s?

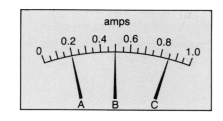

Power supplies and voltage

All electrical circuits need a source of energy. The photographs show some of the supplies you may meet. Each of these provides the **voltage** or **potential difference** needed to make a current flow in a circuit. (It may help if you think of the voltage as being the 'push' which makes charge flow.)

*A normal dry cell gives about 1.5 V. A number of cells can be joined together to make a **battery** which gives a bigger voltage.*

A lead-acid cell gives about 2 V. A car battery uses six cells to give a total of 12 V between its terminals.

This laboratory supply plugs into the mains. It gives a low voltage (up to 12 V) from its terminals.

Measuring voltage

Voltage or potential difference can be measured using a voltmeter.

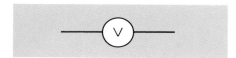

symbol

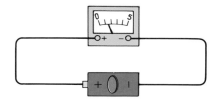

Here a voltmeter is connected across one dry cell. It reads 1.5 V (volts).

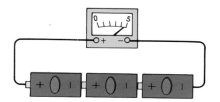

Here the voltmeter is across a battery of three cells. It shows that there is a voltage of 4.5 V.

Unlike ammeters, voltmeters are **not** connected in the main circuit. They are added **across** the component where we want to find the voltage.

Like ammeters, the voltmeters must be connected the right way around. If not, the pointer moves the wrong way and the meter may be damaged.

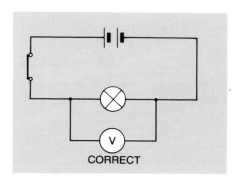

CORRECT

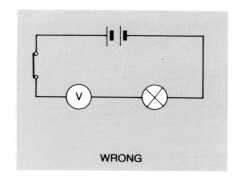

WRONG

Voltages around a circuit

The battery or power supply provides the total voltage for the circuit. In a simple circuit, the voltages across all the components add up to the supply voltage.

The circuits opposite show how the voltage is 'shared out' when all the components, in this case lamps, are identical.

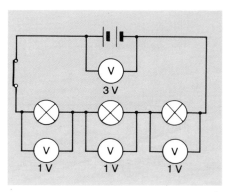

1 V + 1 V + 1 V = 3 V

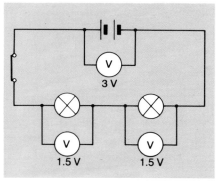

1.5 V + 1.5 V = 3 V

Voltage, current, and resistance

In a circuit the power supply provides the voltage which makes the current flow. Larger voltages cause larger currents.

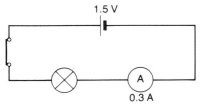

In this circuit 1.5 V gives a current of 0.3 A.

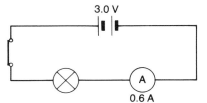

Twice as much voltage (3.0 V) gives twice as much current (0.6 A).

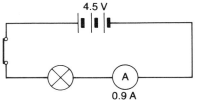

Three times the voltage (4.5 V) gives three times the current (0.9 A).

A battery does not always give the same current. It depends on what is in the circuit connected to it. Some components allow electricity to pass 'easily'. We say that these have **low resistance**. Other components make it 'hard' for current to flow. We say that these have **high resistance**.

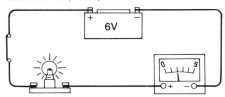

When connected to a 6 V battery, this lamp allows a current of 2 A.

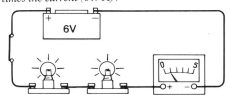

When two of these lamps are connected the resistance is higher and so the current is about 1 A.

Resistances and wires

The wires used for connecting up circuits in the laboratory or in the home have very low resistance. This is because they are made of copper; a metal which is a very good conductor.

The resistance of any wire depends on three main things: its **length**, its **diameter** or **area**, and the **material** from which it is made.

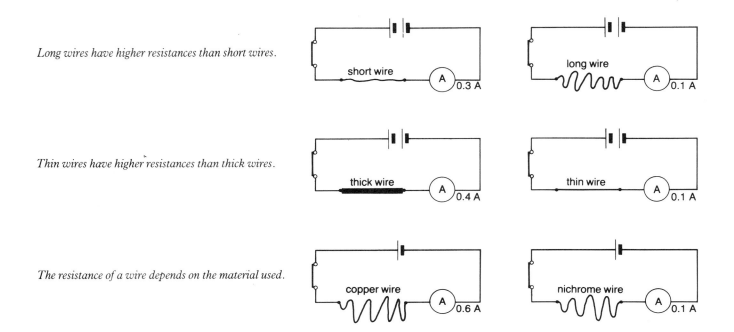

Long wires have higher resistances than short wires.

Thin wires have higher resistances than thick wires.

The resistance of a wire depends on the material used.

Measuring resistance

The resistance of a component or a circuit tells us how many volts we need to apply to make a current of 1 A flow. If a circuit has a large resistance we have to apply a high voltage to get a current of 1 A to flow. If the circuit has a low resistance only a small voltage is needed.

Resistance is measured in units called **ohms**. The symbol for the ohm is Ω (the Greek letter 'omega').

1 V is needed to make a current of 1 A flow through a resistance of 1 Ω.

The table opposite shows the pattern between voltage, current and resistance.

The diagram opposite shows a circuit to measure the resistance of a piece of wire.

The voltmeter is measuring the voltage **across** the wire and the ammeter is measuring the current **in** the wire.

To find the resistance we just divide the voltage by the current.

voltage		resistance		current
1 V	across	1 Ω	gives	1 A
2 V	across	1 Ω	gives	2 A
1 V	across	2 Ω	gives	$\frac{1}{2}$ A
2 V	across	2 Ω	gives	1 A
6 V	across	2 Ω	gives	3 A

The rule is VOLTAGE = CURRENT × RESISTANCE
in symbols $V = I R$

or $\dfrac{\text{VOLTAGE}}{\text{CURRENT}} = \text{RESISTANCE}$ $\dfrac{V}{I} = R$

This is Ohm's law

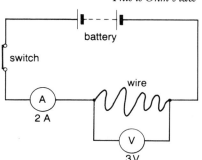

Calculation

Voltmeter reading = 3 V
Ammeter reading = 2 A

$$\text{resistance} = \frac{\text{voltage}}{\text{current}}$$

$$\text{resistance} = \frac{3\,\text{V}}{2\,\text{A}} = 1.5\,\Omega$$

The resistance of the wire is 1.5 Ω.

Questions

1 The diagram shows the circuit set up by a boy trying to measure the resistance of a light bulb. He has made several mistakes. What are they? Draw the correct circuit.

2 When the circuit is set up correctly, it is found that a voltage of 3 V gives a current of 0.5 A.
What voltage would be needed to give a current of 1 A?
What is the resistance of the bulb in the circuit?

3 A girl uses the apparatus to measure the resistance of a copper connecting wire. She finds that 0.1 V (one tenth of a volt) gives a current of 1 A.
   a) What is the resistance of the copper wire?
   b) Why is the girl's value for the resistance likely to be inaccurate? (*Hint*: think of the voltmeter scale!)

Controlling the current

When we need to control the current in a circuit, we can use a **resistor**. This shows a resistor and its standard symbol.

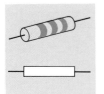

The circuit opposite shows a resistor being used to protect a lamp which would 'blow' if connected directly to the battery. Having extra resistance in the circuit limits the current.

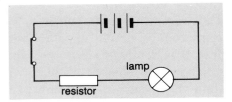

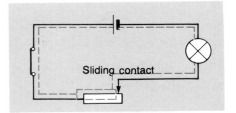

It is often useful to be able to change the current in a circuit. This can be done by using a **variable resistor**. The photograph shows one with a sliding contact being used. As the contact is moved, the length of the resistance wire connected in the circuit changes and so the current changes.

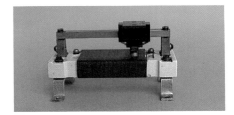

A **perfect** resistor always has the same resistance. The diagram shows a circuit used to see how the current through a resistor changes as the voltage across it is changed. The results are shown in the table and also on the graph.

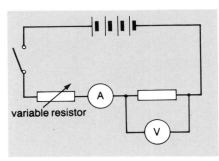

voltage across resistor	current in resistor
0 V	0.0 A
1 V	0.2 A
2 V	0.4 A
3 V	0.6 A
4 V	0.8 A

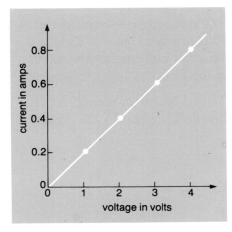

Current-voltage relation for a resistor

Notice that as the voltage doubles from 1 V to 2 V, the current doubles from 0.2 A to 0.4 A. Then as the voltage doubles again (2 V to 4 V) the current doubles (0.4 to 0.8). We say that **the current is proportional to the voltage**.

If we repeat the experiment with a filament lamp in place of the resistor, we find that it does **not** always have the same resistance.

The table of results and the graph show that as the current increases the resistance of the filament in the bulb increases. This is because it is getting hotter! **A wire has a higher resistance when it is hot than when it is cold**.

voltage across lamp	current in lamp
0 V	0.0 A
1 V	0.2 A
2 V	0.35 A
3 V	0.5 A
4 V	0.6 A

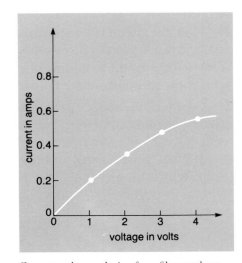

Current-voltage relation for a filament lamp

Series circuits

In the simplest circuits all the components are joined together in one pathway. These are called **series** circuits. There are no branches or junctions in a series circuit.

The diagram shows two cells (batteries), one lamp and three ammeters connected in series. Notice that the ammeters all show the same value (0.5 A).

The current is the same at all places in a series circuit.

In a series circuit, the current has to flow through each component. To find the resistance of the complete circuit we just add up the resistances of all the components.

In the circuit shown, the cells are 'pushing' the current through the 1 Ω resistor **and** the 2 Ω resistor **and** the 3 Ω resistor. The current is exactly the same as if they were connected to one 6 Ω resistor.

Sample question

The diagram opposite shows a series circuit.
Question Why is it called a **series** circuit?
Answer Because the cells and resistors are joined in one pathway which has no branches or junctions.
Question What is the total resistance in the circuit?
Answer Because it is a series circuit we can add the resistances.

$$1\,\Omega + 2\,\Omega + 1\,\Omega = 4\,\Omega$$

Total resistance = 4 Ω

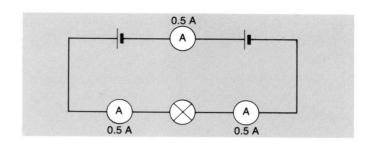

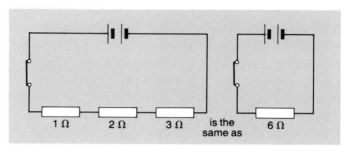

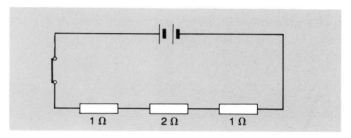

Question If the two cells together give 4 V what is the current in the circuit?

Answer

$$\text{Current} = \frac{\text{voltage}}{\text{resistance}} = \frac{4\,\text{V}}{4\,\Omega} = 1\,\text{A}$$

Voltages in series circuits

The current is the same at all places in a series circuit but larger voltages are needed to 'push' the current through the larger resistances. We can show this by measuring the voltage across each component.

We can find two general rules:

1 The voltages across all the components add up to the voltage given by the power supply. (In this case 3 V + 2 V + 1 V = 6 V.)

2 The voltage splits so that the biggest part is across the biggest resistance. (In this case 3 V across the 3 Ω resistor and only 1 V across the 1 Ω resistor.)

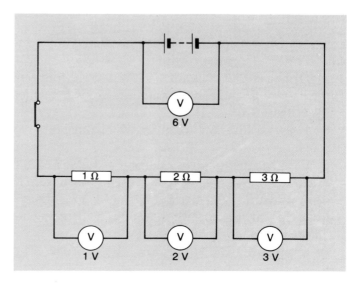

Parallel circuits

Circuits which have junctions where the electrical pathway divides or branches are called **parallel** circuits.

The circuit opposite has two identical lamps connected in parallel. The current 'divides' at junction **X**. Because the two bulbs are the same, the same current (0.2 A) flows through each branch. We can think of the two currents joining again at junction **Y**.

This parallel circuit branches but the top pathway has two bulbs. This means that the top path has a higher resistance and so less current flows through it. This time the current splits at **X** so that the current in the bottom pathway is 0.2 A and the current in the top pathway is just 0.1 A. Notice that the current at point **X** is the same as the current at point **Y**.

This circuit has two, 6 Ω resistors in parallel. The battery gives a voltage of 6 V across each resistor as shown by the voltmeter. The current through each resistor is $\frac{6\,V}{6\,\Omega} = 1\,A$.

The total current in the circuit is therefore 2 A.

The 6 V battery gives a current of 2 A so the total resistance of the circuit must be $\frac{6\,V}{2\,A} = 3\,\Omega$.

Two, 6 Ω resistors in parallel are just like one 3 Ω resistor.

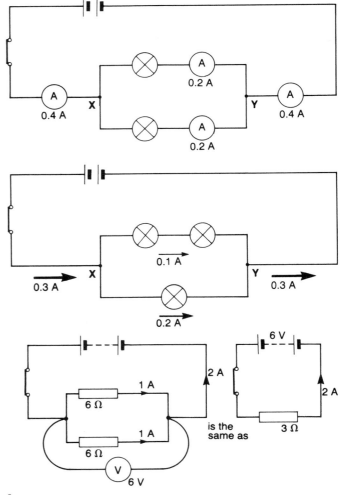

Sample question

A person has three, 30 Ω resistors. How can resistances of **a)** 90 Ω **b)** 10 Ω **c)** 45 Ω be made by joining them together?
The answers are shown opposite.

Answers

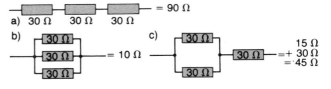

Questions

1 The car shown opposite has two headlamp bulbs and two rear light bulbs connected in parallel with the car's 12 V battery.

   a) Draw a circuit diagram to show how the four bulbs are connected to the battery. Include in your circuit one switch which would control all the bulbs.
   b) Use your diagram to explain why if one bulb 'blows' the other three stay on.
   c) Each headlamp draws a current of 3 A and each rear light bulb draws 0.5 A. What is the total current drawn from the car battery?
   d) A car battery can give a current of 1 A for about 35 hours. A man leaves his car parked with the lights on. Estimate how long it will be before the battery is 'flat'.

Heating effect of an electric current

When a current flows through a wire, some electrical energy is changed to heat and so the wire gets hot. The larger the current, the greater the amount of heat produced.

This heating effect can be dangerous and many house fires are started when faulty wiring overheats. However, it can also be of great use. Many appliances in the home **safely** convert electrical energy to heat.

The diagram shows a light bulb. The filament inside is made from a length of thin tungsten wire. When it is connected to the mains supply a current flows through the filament. Because it has some resistance it heats up. In fact it gets so hot that it glows giving out white light.

The heating element at the bottom of an electric kettle has a length of nichrome resistance wire inside. When it is connected to the mains a current of about 10 A flows through the wire producing enough heat to boil the water.

Of course the copper wires in the lead connecting the kettle to the mains also carry about 10 A because it is a series circuit. However, the copper wires have very low resistance and so very little heat is produced in them. **For the same current, high resistance wires produce more heat than low resistance wires**.

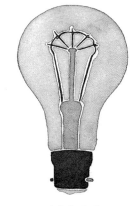

A light bulb

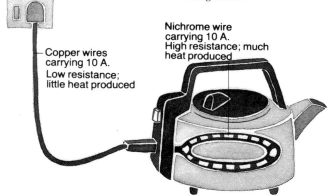

Copper wires carrying 10 A.
Low resistance; little heat produced

Nichrome wire carrying 10 A.
High resistance; much heat produced

Questions

1 List five household appliances which use the heating effect.

2 An electric cooker ring has a heating wire inside a metal tube as shown in the diagram. Why is the packing needed?

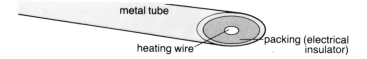

metal tube

heating wire

packing (electrical insulator)

3 A filament bulb is used in a table lamp. Explain why the wire in the bulb gets much hotter than the wires in the lead joining the lamp to the plug.

4 In a hair dryer a fan blows air over heating coils at the front of the dryer. Suggest what might happen if the fan got stuck so that it could not turn.

5 How does an electric blanket work? (Feel one.)

6 The power lines shown opposite are used to transmit electrical energy from one part of the country to another. They carry a current. Why is it important that they have the lowest resistance possible?

Electricity transmission lines

Electricity in the home

Many of the electrical appliances in our homes plug into sockets connected to the mains supply. Mains electricity is produced at power stations by the electricity generating board. It is then sent along power lines to houses, factories, hospitals, schools and other users.

There are two important differences between using the mains as a supply and using a battery:

1 The type of cell or battery used in a torch gives a voltage of about 1.5 V. Mains electricity is at about 240 V!

The mains can also supply very large currents. It is dangerous to touch wires connected to the mains supply. **An electric shock could kill you**.

2 A battery gives a steady voltage and so when it is connected in a circuit the electricity always flows in the same direction. This is **direct current** or d.c.

The mains voltage changes direction very rapidly. This makes the current flow backwards and forwards 50 times in each second. We say that this is an **alternating current** or a.c. with a frequency of 50 Hz. (Hz stands for hertz. 1 Hz is 1 cycle per second.)

The electricity board supplies two wires to your home. One is called the **live** and the other is the **neutral**. The live wire is at about 240 V and the neutral wire is at about 0 V. Electrical circuits in your home are connected to these wires.

This oscilloscope shows the steady voltage from a battery.

This oscilloscope shows how the voltage of the mains supply varies with time.

Why is this dangerous?

Why doesn't this person feel an electric shock?

The Electricity Board's mains supply

Activities

1 Find where the Electricity Board's supply enters your home. If you do not know where this is, ask someone to show you the electricity meter.
See if you can find two thick wires going into the bottom of the meter. One may have red insulation; this is the live wire. The other may have black insulation; this is the neutral wire. **Do not touch these**.

2 Look at a torch bulb and see if you can find its voltage rating. (The value is usually printed on the metal base.) See if you can find the voltage rating of a mains light bulb. (It is usually printed on the top of the glass.)

The ring-main

A light bulb draws less than 0.5 A from the mains supply so the lighting circuit in a house is designed to carry a maximum current of about 5 A. Electric heaters, kettles, and irons take much bigger currents and so must not be connected to the lighting circuit in case they cause overheating. They are connected to separate mains circuits by using plugs in sockets. The sockets in modern houses are connected in a circuit called a **ring-main**.

You should be able to see from the diagram why it is called a ring-main: there is a loop of wire from the live side of the supply right around the house and back again. Similarly there is a ring of wire from the neutral side. Sockets are connected to this loop.

One terminal is joined to the live wire and so is at about 250 V. Another terminal is joined to the neutral wire and so is at 0 V.

Notice that the ring-main has an extra wire. This is called the **earth** wire and is there for safety. How it works is explained later in this chapter.

Adding an extra socket

Extra sockets can be connected to the ring main using a **spur**. The diagram shows how the extra socket is connected. Notice that each socket in the ring-main is connected to the mains supply by two pathways; one around each side of the ring. The spur just uses a single pathway.

Plugs

Appliances are connected to the mains using plugs which have three pins. The longer pin, at the top, makes a connection to the earth wire. As it is pushed in it opens the lower holes in the socket so that its shorter pins can connect with the live and neutral wires.

It is very dangerous to push anything other than a proper plug into a mains socket!

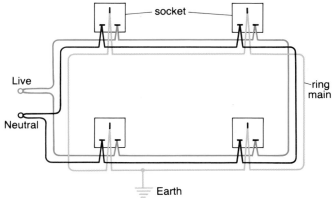

The ring main

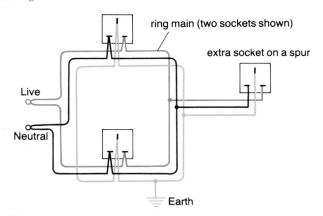

Adding an extra socket

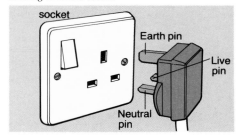

The live and neutral holes are usually covered up for safety. When the long, earth pin is pushed in, the covers are pushed aside so that the neutral and live pins can make contact. It is very dangerous to push anything other than a plug into a mains socket.

Questions

1 The wires used in lighting circuits are thinner than those used in ring-mains. Suggest why this is.

2 Are the sockets in a ring-main connected in series or parallel? Explain your answer.

3 One light bulb draws a current of 0.4 A from the mains supply. A house has eight light bulbs connected to one of its circuits. When all the lights are turned on, what will be the total current drawn from the mains?

Lighting circuits

The lights in a house are usually connected in a **loop** circuit. All the lamps are connected in parallel. Each bulb is connected to the neutral wire on one side and, through a switch, to the live wire on the other. When the switch is closed a bulb has about 250 V across it. This makes a current flow and so the bulb lights.

Two-way switching

In the simple lighting circuit, each bulb is controlled by one switch. Sometimes we want to be able to turn a light on and off from switches at two different places. For example, a light at the top of a staircase needs switches at the top and bottom of the stairs. These must be two-way switches. The circuit symbol for a two-way switch shows that the metal conductor moves between two contacts. Each contact can have a wire connected to it.

The circuit below shows one light bulb being controlled by two switches.

A typical lighting circuit

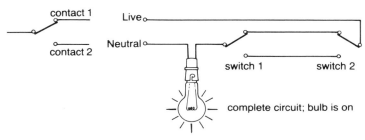

Complete circuit: bulb on

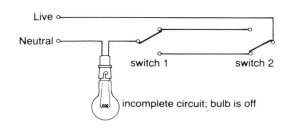

Incomplete circuit: bulb off

Light switches in bathrooms

The type of switch used in a living room would be dangerous in a bathroom. If used with wet hands the water could get behind the switch cover and make a conducting pathway between the live wire and the person. This could give a fatal shock. Pull-switches like the one shown are much safer.

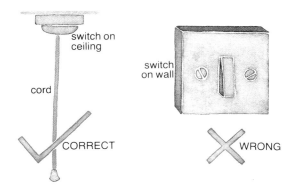

Activities

Carry out a survey in your home to answer these questions:

1   How many light bulbs are used in your home?

2   How many lights are controlled by two-way switches?

3   Why are two-way switches useful   a) on staircases, and   b) in bedrooms.

4   Is the switch in your bathroom a pull-switch? Are there any other rooms in your home where pull-switches are used?

5   Are the light switches in your other rooms made of plastic or metal? Older houses sometimes have brass light switches. Why are plastic ones safer?

Safety and electricity

Our homes would be less comfortable without mains electricity and life would be much harder. However, mains electricity can be dangerous if it is not handled correctly. To prevent damage to property and danger to life the circuits in your home have built in safety measures.

Fuses

When wires carry an electric current they get warm. If the current is too large the wires can get hot enough to start a fire. One way of preventing this is to put a fuse in all the circuits in the home. These are put in the **live** wire close to the electricity meter in a box called the **consumer unit**.

Each fuse contains a thin piece of wire made from a mixture of metals which will melt at a low temperature. If the current gets too high, the fuse wire melts and leaves a gap in the circuit. This stops the current.

Fuses and switches are always connected in the live wire; never in the neutral wire. This is so that they can be used to disconnect the appliance from the 250 V wire. A fuse or switch placed in the neutral wire would work and would stop the current but someone touching the appliance could still get a shock.

When a fuse in the consumer unit 'blows' it must be replaced with one of the correct size. Some can have a new piece of fuse wire fitted but this too must be of the correct rating; 5 A for lighting, 30 A for ring-mains and 45 A for cooker circuits. **It is very dangerous to use anything other than the correct fuse wire**. If a piece of ordinary copper wire is used it will let very large currents flow and may allow a fire to start.

This consumer unit is fitted with fuses. These protect the home's circuits.

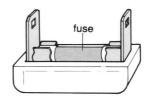 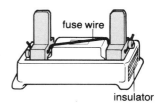

Fuse carrier with cartridge fuse

Fuse carrier with fuse wire (designs may vary)

Miniature circuit breakers

Replacing a fuse or a piece of fuse wire is not easy particularly if all the lights have gone out! Some houses have their circuits protected by small electromagnetic switches called **miniature circuit breakers** or mcb's for short. Normally the switch of the mcb is pushed in to complete the circuit. However, if the current gets too high, the switch springs out breaking the circuit. Once the reason for the circuit overload has been found the switch of the mcb can just be pushed back in until it clicks into place.

Miniature circuit breakers, just like fuses, have to operate at different currents. A 5 A circuit breaker is used to protect the lighting circuit and a 30 A mcb protects the ring-main.

Miniature circuit breakers

Wiring a plug

In Britain, appliances are connected to the ring main using three-pin plugs. The standard plug has square pins; two short pins and one longer one. The wires and pins must always be connected in the same way. To help, the wires in the connecting lead are colour coded. The code should be remembered;

brown = live   **blue = neutral**   **green/yellow = earth**

The tools needed to wire a plug are a small screwdriver with an insulating handle, wire-strippers and a handyman's knife.

Step 1 *Unscrew the large screw which is between the pins of the plug. Remove the lid.*

Step 2 *Loosen the cable clamp by undoing the small screws. It is a good idea to take one screw out completely. Take the fuse out of the plug. This will make wiring the plug easier.*

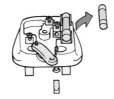

Step 3 *Using a sharp knife, make a slit about 4 cm long in the outer covering of the cable. Pull the wires out of the slit and then cut off the loose bit of covering. Take great care not to damage the insulation of the wires.*

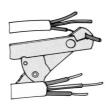

Step 4 *Using wire-strippers pull about 2 cm of the insulation from each of the wires.*

Step 5 *Put the cable under the clamp. Check that each wire will reach its terminal. Do up the clamp so that it grips the cable firmly. Connect each wire to its correct terminal. Make sure that the terminal screw is tightened on to the bare wire.*

Step 6 *Check the following:*
Have you tightened the cable grip? Have you connected the wires correctly according to the colour code?
Have you tightened the terminal screws? Have you made sure that no bare wires are touching?

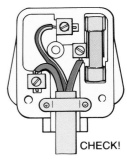

CHECK!

Step 7 *Push the fuse into the holder. Screw on the lid of the plug.*
Test the appliance by plugging it in and switching on.

Note: *the cable for some appliances does not have an earth wire. In this case the larger pin is left unconnected. The two wires must still be connected correctly: brown = live, blue = neutral.*

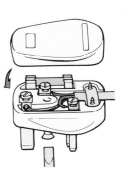

Earthing

If you touch a live wire at about 250 V an electric current can pass through your body to the ground which is at 0 V. The pain you feel is an electric shock. The small current which runs through your body may be enough to kill you even though it is too small to blow the circuit's fuse. Ring-mains and lighting circuits have an **earth wire** to protect users.

Outside each house, close to where the main electricity cables enter, a metal plate is buried in the ground. This is joined to a connector block in the consumer unit where the earth wires from all the house circuits are connected. If a fault lets the live wire become connected to the earth, a large current flows in the earth wire. This makes a fuse blow, often with a bang, cutting off the electricity supply.

Residual current circuit breakers

Sometimes a fault occurs which makes the outside of a properly earthed appliance become live but where the current in the earth wire is not enough to blow a fuse. This could be very dangerous. A residual current circuit breaker can be used to protect against this. It contains an electro-magnetic switch which quickly turns off the power if the current in the earth wire is bigger than 0.03 A. Circuit breakers like this are recommended for use with lawn mowers and portable tools like electric drills. Ask your teacher whether the school laboratories are protected by earth leakage circuit breakers.

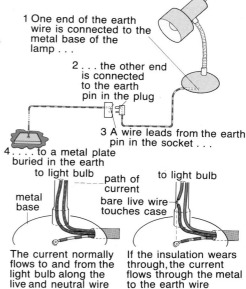

1 One end of the earth wire is connected to the metal base of the lamp . . .

2 . . . the other end is connected to the earth pin in the plug

3 A wire leads from the earth pin in the socket . . .

4 . . . to a metal plate buried in the earth

to light bulb — metal base — path of current

to light bulb — bare live wire touches case

The current normally flows to and from the light bulb along the live and neutral wire

If the insulation wears through, the current flows through the metal to the earth wire

No earth wire!

Appliances marked '**double insulated**' do not have an earth wire. This is because their metal parts are covered with insulating plastic. The live wire can never touch the casing and so cannot give the user a shock.

The diagram shows how the wires should be connected in the plug of a 'double insulated' appliance.

Questions

1 What colour is the insulation of the earth wire?

2 The outside of a hair dryer is made from plastic. Why doesn't it need an earth wire?

3 Explain how the earth wire **and** fuse help to prevent electric shocks in the case of accidents.

4 Why must the fuse be in the **live** wire and not the neutral wire? (*Hint*: think of the voltages of the two wires.)

5 Try to find out the correct first aid treatment for someone suffering from an electric shock.

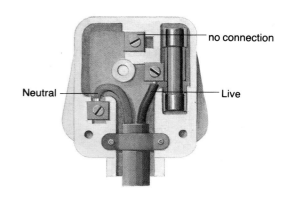

no connection

Neutral

Live

Fuses for appliances

The main fuses in the consumer unit protect the house circuits from damage due to **overload** (too much current). Each appliance is also protected by a fuse in its plug. The diagrams opposite show the type of cartridge fuse used in mains plugs. Inside there is a thin wire made of a metal alloy which melts at a low temperature. If the current gets too great, the fuse wire melts and breaks the circuit. When this happens we say that the fuse has 'blown'.

Fuses can be bought which 'blow' at different currents. They are marked with the maximum current that they can carry for a long time.

Choosing the right fuse

The fuse used in a plug should have a current rating just big enough to let the appliance work properly. If it is rated at too high a current it could allow an overload to cause damage without melting and 'blowing'. Fuses are made in various values – 1A, 2A . . . Household appliance fuses are usually chosen from the following values – 3A, 5A, 10A, 13A.

Calculating currents

Appliances are often marked with their power rating. This tells us how much electrical energy they convert in each second.

$$\text{electrical power} = \text{voltage} \times \text{current}$$

or in symbols: $W = VI$

This gives us a way of working out the current.

$$\text{current} = \frac{\text{power}}{\text{voltage}} \quad \text{or} \quad I = \frac{W}{V}$$

Let us try this for some household appliances where the mains supply is at 250 V. (In the United Kingdom the mains supply is at 240 V. Using 250 V in our calculations makes them much easier and does not make much difference to the answers!)

3000 W electric fire: current $= \dfrac{3000\,W}{250\,V} = 12\,A$

The next highest fuse rating is 13 A so we fit a 13 A fuse in the plug of an electric fire.

100 W table lamp: current $= \dfrac{100\,W}{250\,V} = 0.4\,A$

The next highest fuse rating is 3 A so we fit a 3 A fuse in the plug of a lamp.

Appliance	Current	Fuse
150W stereo 500W food mixer	$\dfrac{150W}{250V} = 0.6A$ $\dfrac{500W}{250V} = 2A$	**3A** (RED)
850W toaster 1000W iron	$\dfrac{850W}{250V} = 3.4A$ $\dfrac{1000W}{250V} = 4A$	**5A** (BLACK)
2000W kettle 3000W heater	$\dfrac{2000W}{250V} = 8A$ $\dfrac{3000W}{250V} = 12A$	**13A** (BROWN)

Investigating the current at which a 1 A fuse 'blows'.

The diagram below shows apparatus which can be used to find the current which will cause 1 A fuse wire to melt. Notice that the light bulb must be able to take a current greater than 1 A or the filament may melt before the fuse wire!

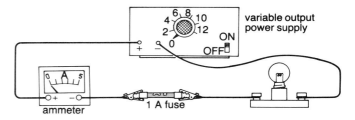

Investigating a fuse

The control on the power supply is set at **zero**. It is then gradually increased. The current rises until the fuse 'blows'. If you watch the ammeter carefully, you can record the maximum current allowed by the fuse. The maximum current is often higher than expected.

Questions

1 What is the purpose of the fuse in
  a) the consumer unit **b)** a mains plug?

2 Why would it be dangerous to use copper wire in a fuse rather than special fuse wire?

3 Suggest why fuses for use in plugs are colour coded.

4 What fuse should you fit in the plug of **a)** a 1500 W sandwich toaster **b)** a 200 W home computer?

Paying for electricity

The electricity board provides each house with a 240 V supply. When we plug an appliance into a mains socket a current flows. In the appliance, electrical energy is changed into another form. For example, a light bulb changes electrical energy into light and heat. **It is the energy which we pay for**.

Energy is measured in units called **joules** (symbol J). The **power** of an electric appliance tells us how many joules of energy it converts in one second. The power is measured in units called **watts** (symbol W).

1 W is 1 J in each second. So, for example, a 100 W light bulb converts 100 J of energy in each second. The power of an appliance used for heating is likely to be much higher and may be measured in **kilowatts** (symbol kW). Remember that 1 kW = 1000 W.

Activities

Try to find the power of the following household appliances: (the easiest way is to use a mail order catalogue) an electric cooker, an electric kettle, an iron, a tumble dryer, a toaster, a vacuum cleaner, an electric drill, a stereo music system and a table lamp.

What do you notice about the power of appliances which contain a heating element compared with the power of those used to produce sound or light?

The longer an appliance is turned on, the more energy it will 'use'. A 1 kW heater switched on for one hour will use 3 600 000 J of energy! (1000 W × 3600 s.) Such numbers are very difficult to work with so the electricity board calculates the energy used in 'units' called kilowatt hours (symbol kW h).

A 1 kW appliance turned on for 1 h will convert 1 kW h of energy. 1 kW h of energy costs between 5p and 6p.

It is easy to work out how much an appliance costs to use:

first   **energy used = power × time**
           (kW h)   (kW)   (h)

then   **cost = energy used × cost of 1 kW h**

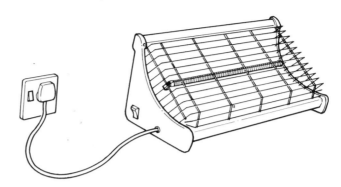

This 1 kW electric fire 'uses' 1000 J of energy in each second

The table below shows roughly how long one unit of electricity will last when using different appliances.

appliance	time to use 1 unit (approximate)
fan heater	$\frac{1}{2}$ hour
steam iron	1 hour
hair dryer	2 hours
power drill	4 hours
colour television	7 hours
freezer	12 hours
table lamp	15 hours

Sample questions

1   If 1 kW h of energy costs 6p, how much does it cost to use a 2 kW heater for 4 hours?

energy used = 2 kW × 4 h = 8 kW h
cost = (8 × 6)p = 48 p

The heater costs 48p to use for 4 hours.

This calculation only works if you remember to use the power in kilowatts and the time in hours!

2   If 1 kW h of energy costs 6p, how much does it cost to use a 100 W light bulb for 5 hours?

energy used = 0.1 kW × 5 h = 0.5 kW h
cost = (0.5 × 6)p = 3p

The light bulb costs 3p for 5 hours use.

Paying the bill

The amount of electrical energy used in the home is measured by a meter. This is connected by the consumer unit. It is sealed with a special lead tag so that it cannot be tampered with.

When current flows through the meter the digits or numbered dials (or the pointers) move. These show how much energy has been used. The electricity board reads the meter every three months. A bill, like the one below, is then sent for payment.

A digital meter

Activities

1   Find the electricity meter in your home. Draw a diagram of it showing all the digits or dials clearly. Write down the meter reading.

2   Look at the disc **inside** the meter. It spins around when electricity is being used.

Get someone to turn different appliances on and off while you watch the disc. What happens when an electric fire, or a cooker or an immersion heater is turned on? What happens when a light is turned on?

3   Each morning for two weeks **at the same time** read your electricity meter. Keep a record of your results in a table.

Do you use the same amount of energy each day? If you used a lot more on some days try to explain why.

4   Ask if you can see the last electricity bill for your home. Check that it asked for the correct amount of money. Try to find out what a 'red' bill means.

Questions

Using the electricity bill opposite answer the following questions.

1   What was the meter reading at the start of the three months?

2   What was the meter reading at the end?

3   How many units were used in three months?

4   How much does one unit of electricity cost?

5   How much must be paid for the electricity used?

6   Does the Board charge a fixed charge for the supply?

7   How much must be paid in total?

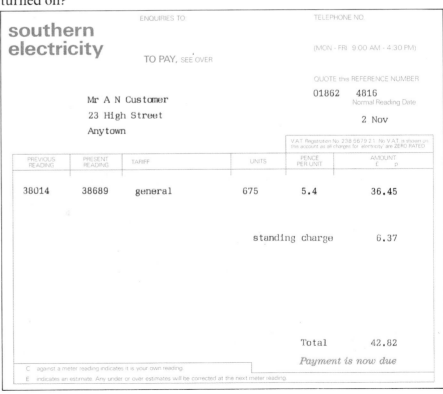

southern electricity

ENQUIRIES TO:

TELEPHONE NO.

TO PAY, SEE OVER

(MON - FRI 9:00 AM - 4:30 PM)

Mr A N Customer
23 High Street
Anytown

QUOTE this REFERENCE NUMBER

01862   4816
Normal Reading Date

2 Nov

V.A.T. Registration No. 238 6679 21. No V.A.T. is shown on this account as all charges for 'electricity' are ZERO RATED

PREVIOUS READING	PRESENT READING	TARIFF	UNITS	PENCE PER UNIT	AMOUNT £ p
38014	38689	general	675	5.4	36.45
		standing charge			6.37
				Total	42.82

Payment is now due

C   against a meter reading indicates it is your own reading.

E   indicates an estimate. Any under or over estimates will be corrected at the next meter reading.

Questions

1 A student uses the apparatus below to test whether materials are electrical conductors or insulators. The material is connected to clips **X** and **Y**.

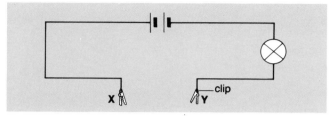

a) What is a conductor?
b) Which of the following would make the bulb light if they were connected between **X** and **Y**; iron wire, glass rod, polythene (plastic), rubber, aluminum foil?
c) When the student puts some copper wire between **X** and **Y** the bulb does not light. Suggest what might be wrong with the apparatus. How could the student check?

2 a) A student connects a bulb rated at 3 V to three 1.5 V cells as shown below.

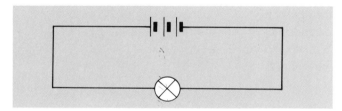

The bulb gives a bright flash and then goes out. Explain why this is.
b) The student then connects a new bulb in the circuit below.

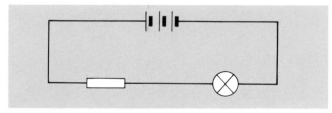

The bulb glows brightly and stays on.
i) What component has the student added?
ii) How does the new component protect the bulb?

3 A science teacher does not have enough variable resistors for a class practical so she uses the apparatus below.

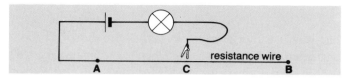

When the clip is attached to the resistance wire at **C** the bulb glows dimly.
a) Which way should the clip be moved to make the bulb glow more brightly?
b) Explain why this arrangement acts as a variable resistor.

4 A student makes a model of a house lighting circuit. The circuit he uses is shown below.

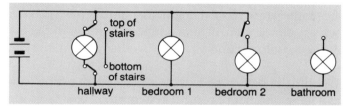

a) Which lamp will always be on?
b) Which lamp will never be on?
c) Explain why the light for the hallway cannot always be turned on and off by the switches at the top and bottom of the stairs.

5 The diagram below shows the plug connected to the flex of a metal table lamp.

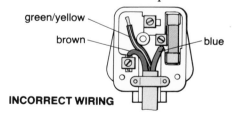

The wiring of this plug is dangerously wrong but the lamp still works.
a) Why does the lamp still work even though the live and neutral wires have been reversed?
b) Because the plug has been wired wrongly the switch is now connected to the neutral side of the supply. Explain why it would be dangerous to change the bulb in this lamp even with the switch turned off.

8 Feedback and control

What do we mean by control?
What type of things need control?
Why do they need to be controlled?

Machines use energy to work. If they are to work properly, they must be **controlled**. For example, the electric motor in a record player must be made to turn at the right speed. If it does not, the music sounds strange. Try playing an album at 45 rpm instead of 33 rpm!

The driver of a car has constantly to control the speed of the engine. If it turns too slowly the engine may stall and the car stop; if it turns too quickly an accident may be caused as the car goes 'out of control'.

Where the 'machine' causes heating, the temperature may need to be controlled. For example, in an oven the food will not cook properly if it is too cold, but will burn if it is too hot.

The temperature of a car engine is controlled by a cooling system. The car uses less petrol at fairly high temperatures, but if the metal parts got too hot they would expand and rub together causing the engine to stop.

The human body is the most complicated 'machine' you will ever study. It has a large number of control systems to keep it working. An example is your balancing system. When you walk, you balance on one foot before moving your weight until you tip over on to the other foot. You then balance there for a very short time before you tip over again. This has to be controlled if you are not to fall over! It is not surprising that engineers have found it difficult to build a robot which can walk like a human.

For living things, we may need to control the conditions that surround them. For plants this means controlling the temperature and the amount of light and water. The photograph shows a terrarium. The glass case traps much of the water lost from leaves and from the soil. This allows plants, like ferns, which prefer moist air to be grown in dry, centrally-heated rooms.

Activities

1 Copy and complete the table below for ten things you would find in the home that need control. The first one is done for you.

What is it?	What is being controlled?	What might happen without any control?
Record Player	Speed record turns	Music would sound wrong!

2 Try to stand on one leg for two minutes. Explain how you keep your balance.

Record deck with speed control

Formula 1 racing cars; controlled power

Learning to control balance!

Terrarium

Manual or automatic control?

People are very good at controlling things; we do it all the time without thinking. For example:

A man sitting in a room decides that it has become too cold. He turns on the electric fire. He is controlling the temperature of the room.

Controlling room temperature . . . manually

This is an example of **manual** control because a person presses a switch to make things change. An automatic system does not need a person to operate it. A mechanical or electrical device is used instead.

An electric toaster is a good example of an **automatic system**. When we cook toast under a grill, we are supposed to turn the grill off when the bread turns a golden-brown colour. However, people are easily distracted and sometimes forget what they are doing. The result in this case is burnt toast! In an automatic toaster, a detector is used to turn off the electricity when the toast is done.

Our modern society uses a great deal of energy and a huge number of machines. Automatic control systems make our lives safer and more comfortable.

The picture opposite shows a lamp lighter. His job was to walk the streets each evening and to light the gas lamps. In the morning, each lamp had to be turned off. The other picture shows a modern road system at night. Each lamp is automatically turned on and off.

A Lamplighter lighting gas lamps

Questions

1 Use a dictionary to find out what *manual* means.

2 What does *automatic* mean?

3 Some cars can be bought in two versions: 'manual' and 'automatic'. What does this mean?

4 Why do we not have lamp lighters any more?

5 What advantages does our modern, automatic lamp lighting system have? Can you think of any disadvantages?

6 List any other jobs that you think an automatic system would do better than a person.

Automatically controlled street lamps

All control systems have sensors. These 'sense' or detect changes in conditions. When conditions change, the sensor changes and this can be used for control. There are many different types of sensor.

Sensing changes in temperature

Many sensors use the fact that things expand when they are heated. For example the mercury in a thermometer expands when it gets hotter. The top of the mercury thread moves up the scale showing that the temperature is rising.

Another useful sensor is the **bimetallic strip**. This is made of two metal strips of equal length fixed together. Usually one piece is brass (or copper) and the other iron. At room temperature the strip is flat. When it is heated the two metals expand but the brass expands more than the iron. This makes the bimetallic strip bend. If a bimetal strip is cooled down the brass contracts more than the iron and so the strip bends the opposite way.

The bimetallic strip is useful for automatic control because it is made of metal and can be used as a switch in electrical circuits. The circuit opposite is a simple fire alarm. When the bimetallic strip is cold, there is a gap in the circuit. When the strip gets hot, as in a fire, the brass expands bending the strip upwards. If the temperature gets hot enough, the strip touches the metal contact and completes the circuit. This makes the bell ring.

Thermistors

The **thermistor** is another useful temperature sensor. It is an electrical component which, when cold, makes it difficult for electricity to pass but, when hot, allows larger currents. We say that a thermistor has a high resistance when cold and a low resistance when hot.

The circuit symbol for a thermistor is:

The engine of a car must not be allowed to get too hot. There are control systems to prevent this but there is also a temperature gauge to tell the driver whether the temperature is too high (or too low). Usually the sensor is a thermistor.

Activities

1 The words thermometer and thermistor start with 'therm'. What do you think '*therm*' means? Use a dictionary to list other words which start with 'therm' and find their meanings.

2 Find the engine temperature gauge on a car dashboard. Draw the face of the gauge. Mark on the low, high and normal temperature positions.

3 Look at the other gauges on a car dashboard. What types of sensors do you think they are connected to?

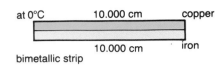

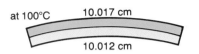

A bimetallic strip before heating

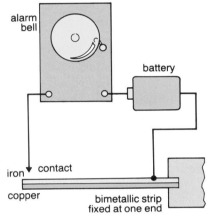

The strip bends because the copper expands more than the iron.

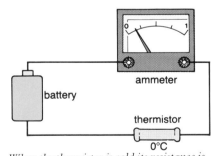

Design for a fire alarm

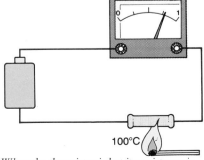

When the thermistor is cold its resistance is high: little current flows.

When the thermistor is hot its resistance is low: a larger current flows.

More sensors

It is not only changes in temperature which we need to detect. We need other sensors if we are to control a large variety of things. Some useful sensors are shown below.

Light sensor

Automatic lighting circuits need a sensor which can detect changes in light intensity (brightness). Many use an electrical component called a **light dependent resistor** (**LDR**). The resistance of the LDR is high when it is dark and low when it is in bright light.

Proximity sensor

Some control circuits need a sensor to detect when things are in position, for example when a door is closed. The simplest sensor is a reed switch. This has two parts; a switch consisting of two strips of metal arranged so they are not quite touching, and a magnet. When the magnet is close to the switch, the two strips of metal touch. This can be used to control an electrical circuit.

Liquid level sensor

Many machines contain liquids. A car contains water to cool the engine and to wash the windscreen, hydraulic fluid to work the brakes and, of course, petrol. The level of the liquid can be sensed in a number of ways. A petrol tank usually has a float which moves the contact on a variable resistor. This controls the current in a circuit. A meter in the dashboard of the car gives a high reading when the petrol level is high and a low reading when the car is low on petrol.

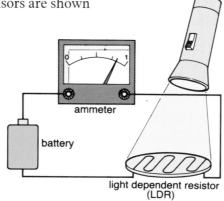

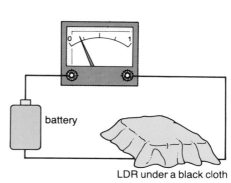

The LDR has a low resistance in bright light: the ammeter shows a large current.

The LDR has a high resistance in the dark: the ammeter shows a smaller current.

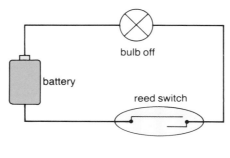

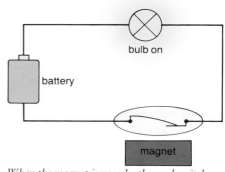

When the magnet is not near by the reed switch is open.

When the magnet is near by the reed switch closes.

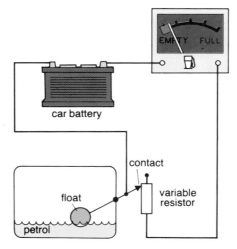

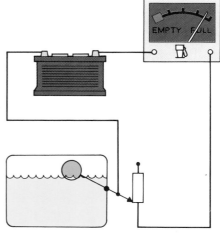

When the petrol is low the float is low. This moves a contact so that there is a large resistance in the circuit. The meter shows a small reading.

When the petrol is high the float is high. This moves the contact to give a small resistance in the circuit. The meter shows a large reading.

Coordination and control in the body

Humans are sensitive animals. To survive we need to be able to sense change in our surroundings. We can detect things like sudden noises, flashes of light, changes in temperature, and pressure on the skin. We call these **stimuli**.

We need to be able to respond quickly to these things. For example, if someone on the hockey field shouts 'duck!' we may have to move suddenly to avoid being hit by the ball. This reaction is coordinated by our **central nervous system**. The central nervous system is made up of the brain and spinal cord. It is the control unit for all our actions.

The brain and spinal cord receive messages from sensors throughout the body. These sensors are called **receptors** and include our skin, our eyes, nose and taste buds. The messages pass along nerves to the central nervous system. This then sends messages along other nerves to muscles and other **effectors** to make them respond. The nerves are made up of special cells called neurones. **Sensory neurones** carry messages from receptors to the central nervous system. **Motor neurones** carry messages from the central nervous system to the muscles and other effectors.

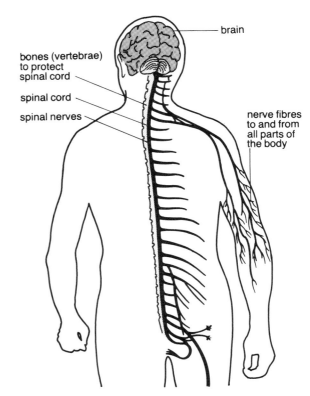

The central nervous system

Reflex actions

Sometimes animals have to react very quickly if they are not to be injured or killed. Such actions are controlled by the central nervous system and are called **reflex actions**. These are automatic responses which you cannot consciously control.

stimulus	reflex action
flash of bright light	pupil of eye gets smaller
speck of dust touches eye	eye blinks
food enters wind pipe ('goes down the wrong way')	coughing
hand touches 'live' wire (electric shock)	arm moves away quickly

All these reflex actions are designed to keep you safe. There is no time to think about what to do. The body protects itself by reacting automatically. The example in the diagram shows what happens when you touch something which is very hot.

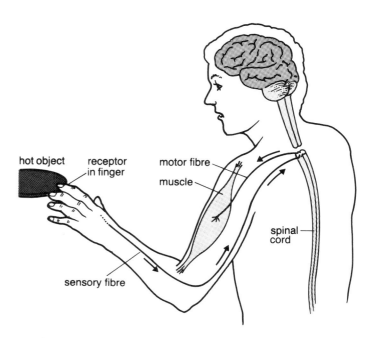

A reflex action

1 *Finger touches hot object*

2 *Receptor detects high temperature*

3 *Message sent to spinal cord through sensory fibre*

4 *Spinal cord sends message along motor fibre*

5 *Effector (muscle) responds to pull finger away*

Reflex action and the brain

The reflex action described on page 158 has a pathway of nerves, or **reflex arc**, which acts through the spinal cord. This is known as a spinal reflex. Some other reflex actions act through the brain. In Activity 1 you will see the knee-jerk reflex. The nervous pathway for this is through the spinal cord. The pupil reflex action seen in Activity 2 has a nervous pathway through the brain.

Voluntary actions

Reflex actions are extremely quick and are designed to keep us from danger. However, humans live very complicated lives and respond intelligently to a wide variety of situations. Reactions which we think about and can consciously control are called **voluntary actions**.

Voluntary actions are controlled by the brain. For each action, impulses are sent to the brain from a large number of sensors. These are considered in the brain before messages are sent to the muscles and other effectors. For example, if you pick up a very expensive dinner plate and find it to be hot your reflex action would be to drop it immediately. However, your brain gets impulses from light sensors in the eye telling it that this is an expensive plate. It will also receive impulses from the spinal cord telling it just how hot the plate is. It will quickly consider these and if your fingers are not being damaged too much, the brain will send impulses to the muscles of your hand and arm so that you put the plate down carefully. This example has been greatly simplified.

The brain is a very complicated organ. We can consider it to be the control centre for voluntary actions. Certain parts of the brain control different actions. Conscious thought and memory take place in a part of the brain called the **cerebellum**.

Questions

1 What is the purpose of the central nervous system in humans?

2 Why do you think the spinal cord runs through the middle of the bones (vertebrae) of your backbone?

3 Your skin contains 'pain' receptors. Draw a diagram to show what happens when you prick yourself with a needle.

Activities

1 Sit on a chair and cross one leg over the other.

Get someone to tap you sharply just below the knee with the edge of a ruler. Do not watch.

What happens to your leg?

Can you stop this from happening?

Swap places and try the test on your partner. Is his or her reaction the same? Is there anyone in the class who does not react in the same way?

2 Get someone to sit facing you. Look closely at his or her eyes and then slowly move a beam of light from a torch across your partner's eyes.

What happens to the pupil of the eye when in bright light?

What happens to the pupil of the eye as it gets darker?

(Some people react badly to having bright light in their eyes. Do not place the torch too close to your partner's eyes.)

3 If you want to test yourself, set a lamp to shine into a mirror. Look into the mirror and then cover *one* eye with your hand. As you remove your hand watch what happens to your pupil.

Just out of interest, watch what happens to the pupil of the eye which remains open when you cover and uncover the other eye.

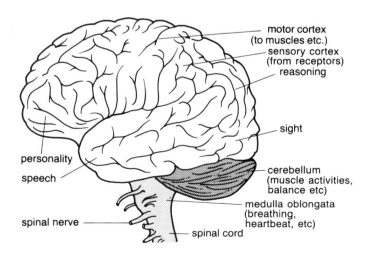

The brain and its control centres

Keeping things steady

Sometimes it is important to keep conditions constant. In an oven, for example, a cake may need to be cooked at a temperature of 200°C for two hours. If the temperature of the oven keeps rising, the cake will burn. We could turn the oven off when the temperature got to 200°C but then, as it cooled down, the cake would not cook properly. What is needed is an automatic control system which will keep the temperature steady. A device which keeps temperature steady is called a **thermostat**. In a gas oven, the thermostat controls the flow of gas to the burners. When the oven is cold gas can pass through the valve into the burner. As the temperature rises the brass tube expands and so end *A* moves to the left. Invar is a metal alloy which does not expand much when heated and so end *B* is pulled to the left shutting off the gas to the burners. If the temperature in the oven starts to fall, the brass tube contracts. This pushes end *B* to the right opening up the valve and increasing the gas supply.

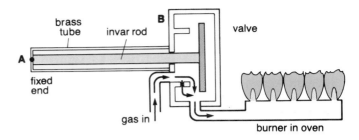

Gas thermostat when the oven is cold

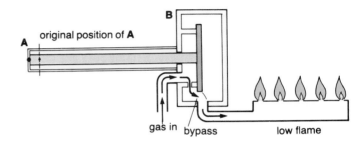

Thermostat when the oven is hot

When ironing clothes the iron has to be at the right temperature. If it gets too hot the material may be scorched or may melt. If the iron gets too cold, the creases will not come out of the material. The electric iron shown in the photograph has a thermostat to keep the temperature constant while the clothes are being ironed. The sensor used in the thermostat is a bimetallic strip (see page 156). When the iron is below its working temperature, the circuit is complete and so the coil heats up. As the temperature rises, the bimetallic strip bends until finally the circuit is broken. Here the iron has reached its working temperature. The contacts have moved apart so the heating coil is turned off.

How do you think the temperature control knob works?

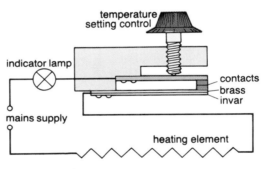

Electric iron thermostat

Thermostatically controlled iron

Activities

1   Care labels on clothes often give the recommended iron temperature using the following diagrams;

Find out what the symbols mean.
Try to find out the approximate temperatures for these settings.
Name some materials which should be ironed at these settings.

2   Investigate the time it takes for an iron to reach its hottest working temperature from cold.
How long does it stay hot before the thermostat switches on again?
(Ask permission first and take care: a hot iron is at more than 200°C!)

Modern central heating systems have thermostats fitted to each radiator. These allow the temperature of each room to be controlled.

The thermostat shown in the photograph contains a flexible metal container filled with gas. As the room gets warmer, the pressure in the container increases. The container expands and pushes the valve down. This stops the flow of hot water into the radiator.

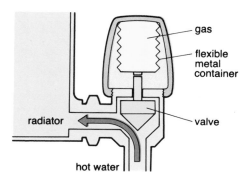

Radiator thermostat

Radiator thermostat

Pressure cooker

In a pressure cooker, water boils at a temperature greater than 100°C. This is because the pressure above the water is more than the pressure of the atmosphere. Food is cooked much more quickly at these higher temperatures. As the pressure cooker is heated, the pressure of the air and water vapour inside increases. If there was no control system, the cooker would eventually explode. To prevent this a simple valve is fitted. Below the operating temperature of the pressure cooker the weight of the valve keeps it in place. Steam cannot escape. As the pressure inside the cooker increases the upward force on the valve gets larger. Eventually the upward force becomes large enough to lift the valve. Steam escapes through the valve. (You may have heard a pressure cooker hissing as this happens.) Because some steam has escaped, the pressure inside the cooker falls slightly and the valve drops back into place. The pressure then starts to rise again.

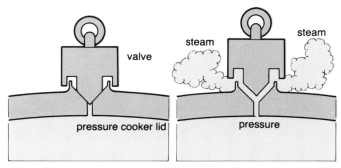

When the pressure cooker is heating up, the weight of the valve keeps it in place. No steam escapes.

When at the operating temperature, the pressure inside the cooker is just enough to lift the valve. Some steam escapes.

Water tank

Most houses have a water tank somewhere in the roof space. A ball-valve, like the one shown in the diagram, is used to keep the water level in the tank constant. The ball floats on the water and the arm holds a valve shut. This stops any water coming into the tank. When water is used in the house, the water level in the tank drops and the ball moves downwards. This opens the valve and so water from the mains pipe pours into the tank until the correct level is reached.

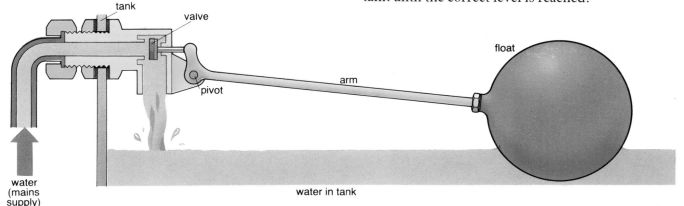

Ball-valve for a water tank

Feedback

Control systems for keeping things steady have one thing in common; when the sensor detects a change it activates the system to bring about a change '*in the opposite direction*'. For example, when the iron **cools down** the thermostat turns the heating element on to make the iron **heat up**. When the **water level falls** in a cistern, the ball-valve opens to make the **water level rise** again.

We say that these systems use **negative feedback.**

Tight-rope walking needs control!

Control systems normally use negative feedback. **Positive feedback** is where the output of a device, rather than keeping things steady, causes conditions to change even more! You may have heard the loud screeching noise when a microphone is placed too close to the loudspeaker of an amplifier system.

In this example of positive feedback, a small input signal is amplified but then the output (loud sound) is fed back to the sensor. This gives a larger input which is amplified and so on until the amplifier is working at its limit and giving a loud screeching noise.

Positive feedback is used in some alarm systems but negative feedback is much more useful for control.

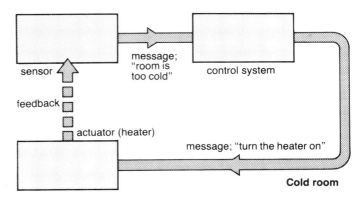

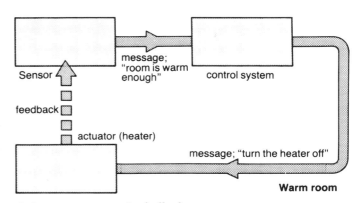

A thermostat uses negative feedback

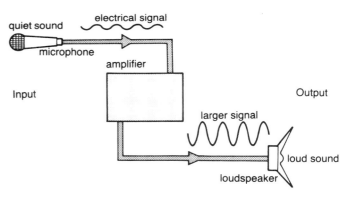

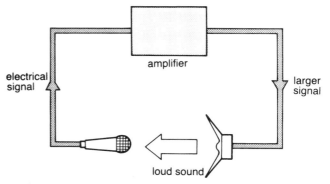

A system with positive feedback

162

Controlling a micro-environment

The space in which we live is called the environment (see Chapter 6). We sometimes think of the whole planet as being our environment. However, we can think of our homes or even one room of a house as a **micro-environment**. Notice that it is easy to control the temperature of a room but it would be difficult to have any control over the temperature of the whole atmosphere.

If we take plants or animals from their natural environments, they may die unless we keep them somewhere where the conditions are controlled. The photograph shows a large greenhouse used for growing tropical plants. The plants would die in the cold, wet conditions of England but inside the greenhouse the temperature and the amount of water can be carefully controlled. As a result, even banana trees grow well.

Palm House, Kew

Control in an aquarium

Tropical fish live in the warm rivers and oceans of tropical regions. The fish are often brightly coloured and very attractive. For this reason some people like to keep them in an aquarium in their room. The conditions in the aquarium have to be just right at all times or the fish will die.

The photograph shows a well-kept fish tank. It has a heater controlled by a thermostat. This keeps the water at the correct temperature. Tropical fish need to be kept at about 25 °C. If they get too cold or too hot they will die. The heater is controlled by a bimetallic strip just like the one used to control an electric iron (see page 160).

A carefully controlled aquarium

Fish use oxygen dissolved in water for respiration. In a river or lake, oxygen from the air can dissolve in the water over the large surface area. In a tank the surface area is much smaller and so there is a danger that the oxygen level will become too low and the fish will die. To prevent this a pump is used to pump air through a tube to the bottom of the tank. The rising bubbles stir up the surface of the water allowing more oxygen to dissolve.

Questions

1   a) How are plants in a greenhouse protected from very low temperatures?
b) How can plants in a greenhouse be protected from very high temperatures?
c) There are several different kinds of automatic watering systems for greenhouses. *Either* find out how one of these works *or* try to design your own system.

2   The graph opposite shows how the water temperature in an aquarium varies.
a) What is the average temperature of the water?
b) What is the maximum water temperature?
c) What is happening to the bimetallic strip in the thermostat when the water temperature is at a minimum?

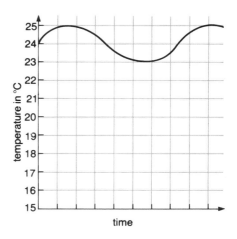
The variation of temperature in an aquarium

Keeping things steady in the body

Machines can usually work under a wide range of conditions. Living things are much more sensitive. If the conditions vary too much the animal or plant will die.

Our bodies have large numbers of control systems. These keep our internal conditions steady and just right for living. This is called **homeostasis**.

Keeping the body temperature steady

A doctor can usually tell if a patient is ill by taking the body temperature. This is because all healthy humans have a body temperature close to the average of 37 °C. Only disease or some other disorder can make it vary. A change of just 2 °C can mean that something is seriously wrong. The human body uses a number of control systems to keep its internal temperature steady.

The body can raise its temperature using energy from two sources: from chemical reactions taking place inside cells (for example, the contracting cells of muscles release energy); from the surroundings. For example, energy from the sun will be absorbed by our skin. We can also gain energy from hot food or drink.

The body loses energy from the skin by radiation and conduction. Evaporation from the skin also causes heat loss. The air that we breathe out is much warmer than the air which we breathe in. This means that we lose some energy from our bodies as we breathe.

For most of the time, the energy losses are balanced by the energy produced inside our bodies and that gained from the surroundings. The body temperature therefore stays at 37 °C. If the surroundings heat up or cool down or if a person does strenuous exercise, then the body's control system has to come into action.

The body's systems are controlled by the brain. Blood passing through part of the brain called the **hypothalamus** has its temperature checked. If the temperature is not correct, the brain sends 'information' to parts of the body which react to control the body temperature. The skin plays a very important part in keeping the body temperature steady.

Overheating

If the brain detects a rise in blood temperature it makes changes occur in the skin. Firstly the small blood vessels in the skin, called capillaries, become wider. This allows more warm blood to flow close to the surface of the skin and so lose more heat energy. You may have noticed how people look red in the face after exercise. This is because the blood vessels near the surface of their skin are carrying more blood. This way of losing heat is called **vaso-dilation**.

Secondly, the sweat glands pour sweat on to the surface of the skin. This thin layer of liquid evaporates taking energy from the body. This has a cooling effect.

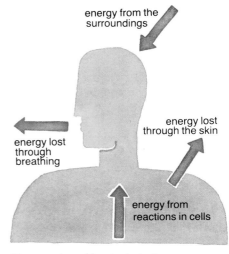

Energy gain and loss in the body

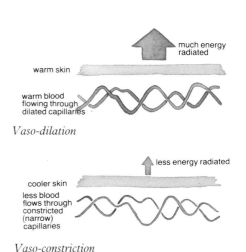

Vaso-dilation

Vaso-constriction

Insulation and the body

Overcooling

If the body loses energy faster than it can produce it the effects can be very serious. In cold weather for example, old people without enough hot food or warm clothing lose too much energy. This makes their internal body temperature fall to dangerous levels. This is called **hypothermia**. If the body temperature becomes too low the person will die.

Normally the body's control system tries to overcome this in four main ways.

1   Sweat production stops. The surface of the skin becomes dry and so no energy is lost in evaporation.

2   The blood vessels near the surface of the skin become narrower. This means that less warm blood flows close to the surface and so less energy is lost by conduction. The narrowing of the blood vessels is called **vaso-constriction**.

3   The muscles in the limbs have uncontrollable bursts of contraction and relaxation. We feel this as **shivering**. The chemical changes in the muscles release energy and this raises our body temperature slightly.

4   Like all other mammals our skin is covered in hairs. Each hair has a small muscle attached to it just inside the skin. When it is cold, the muscle is made to contract and this pulls the hair upright. The hairs then trap a layer of air around the body which acts as **insulation**.

This works well in mammals with furry coats but does not do much good in humans where the hair is sparse. However, each goose pimple shows where a hair has been pulled into position. Of course we have a better way of insulating ourselves; we put on layers of clothes!

Controlling body temperature in polar regions

Beware hypothermia

Activities

1   Take your body temperature by placing a clinical thermometer under your tongue. Now take your skin temperature as accurately as you can. What difference do you notice?

2   If everyone in the class has taken their body temperature, work out the average. What is the lowest value in your class? What is the highest? What would you say if someone in the class measured their body temperature to be 45°C?

3   Using some cotton wool, rub some alcohol on the back of your hand. (Perfume or aftershave works just as well.) What do you feel? Why?

4   Take your body temperature before and after strenuous exercise. Can you find any difference?

165

Keeping the right amount of water in the body

Our body contains a large amount of water. Much of it is in our blood. We get our water from food and from the liquids we drink. We lose water from our body as we breathe out, when we sweat and when we urinate. The balance of water and salts in our body must be carefully controlled if we are to stay healthy.

The sensor for our water control system is a part of the brain containing the hypothalamus and a small stucture called the **pituitary gland**. As blood flows through the hypothalamus the amount of water is monitored. If the water level is too low, the hypothalamus makes the pituitary gland give out a chemical known as **ADH** (anti-diuretic hormone) into the blood. This acts as a message to the **kidneys** where the water level is controlled.

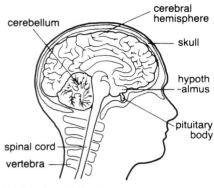

Section through the head

In our bodies we have two kidneys. One of their functions is to control the amount of water in our blood. The photograph shows a sheep's kidney. It is about the same size and shape as those in an adult human.

The outer part of the kidney is called the **cortex**. This part contains very small 'knots' of thin blood capillaries. Each 'knot' is called a **glomerulus**. The inner part of the kidney is called the **medulla**. This part contains about a million tiny tubes called **nephrons**.

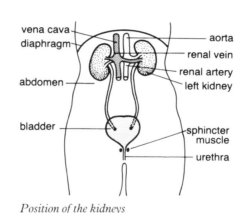

Position of the kidneys

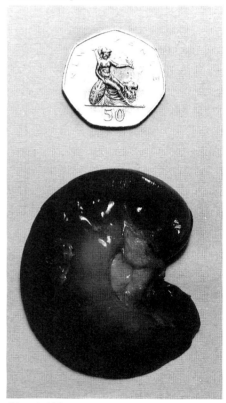

Sheep's kidney

Activities

Get a whole lamb's or pig's kidney from the butcher's shop. If it has fatty tissues around it, carefully remove the fat using your fingers and/or a pair of scissors. Try to find the ureter.

Using a pair of scissors, cut the kidney lengthways so that you finish up with two 'kidney shaped' pieces. Try to find the cortex and the medulla. Look carefully and note any differences in colour and texture between the two regions. Can you find any large vessels (tubes) inside the kidney?

Draw a labelled diagram of the kidney you have dissected.

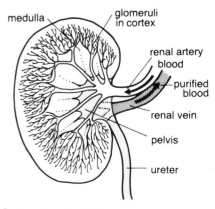

Section through a kidney

Control and the kidneys

Each nephron starts in a cup-shaped structure called **Bowman's capsule**. A glomerulus sits in this cup as shown in the diagram. As blood flows through the capillaries, substances such as water and dissolved salts are filtered as they pass into the Bowman's capsule. Large molecules such as proteins cannot pass through the wall of the capillary and so are left in the blood.

The liquid which does pass into the Bowman's capsule is mainly water and waste materials. Any useful substances, such as glucose, which have got through are reabsorbed by blood vessels. The rest of the liquid passes down the long, looped part of the nephron called the **loop of Henle**. If more water is needed in the body, a lot of it is reabsorbed here by the capillaries of the kidney's main vein; the **renal vein**.

After the necessary water has been reabsorbed, the nephron contains a solution containing waste materials such as **urea**. The solution is called **urine**. This passes through a tube called the **ureter** to the **bladder**. Urine is stored in the bladder until it is passed out of the body. **Blood leaving the kidney is purified and contains the correct amount of water.** Urine leaving the body carries away unwanted water and any unwanted substances.

The amount of water reabsorbed in the loop of Henle is controlled by the ADH sent from the pituitary gland. If the blood water level is low then the gland sends more ADH which increases reabsorption. Once the water level is correct the brain reduces the level of ADH. This is a form of negative feedback.

Questions

1 Where is the water content of the blood 'detected'?

2 How are messages sent from the brain to the kidneys?

3 What is a capillary? What is a glomerulus?

4 Why can't protein molecules pass from the glomerulus into the Bowman's capsule?

5 If the blood water level is too low, where do the blood vessels reabsorb water from the nephron? What else is reabsorbed here?

6 Why do we say that the body's water control system uses negative feedback?

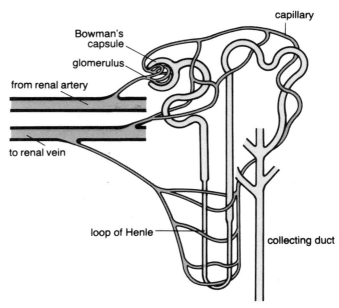

Enlarged view showing the loop of Henle

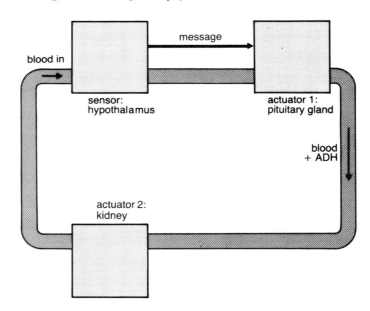

Stage 1 Blood has too little water. This is detected in the brain.
Stage 2 Pituitary gland secretes more ADH to 'tell' the kidneys.
Stage 3 Blood vessels reabsorb more water in the loop of Henle.
Stage 4 Blood from kidneys has correct amount of water.
Stage 5 Brain detects that the water level is correct.
Stage 6 Pituitary gland secretes less ADH.

Electronic control

In recent years our homes, schools, offices and factories have changed as more and more electronic devices have been produced. The most obvious are things like calculators and digital watches. There are, however, other things which have been improved by including electronic control systems. Electronically controlled washing machines, central-heating boilers and burglar alarms are readily available. Modern car engines have electronic control systems to make them perform better and to make them more economical to run. The list of things that can be electronically controlled is almost endless. **All these things use electronically operated switches for control**.

The transistor

We have already seen switches in control circuits. Perhaps the simplest example is the switch which controls the light in a room. This switching can be carried out using a transistor as an electrically operated switch.

The photograph shows a transistor and the diagram shows its symbol. You can see that the transistor has three 'legs' where electrical connections can be made. These are shown by the three lines on the symbol. These connections are known as the **base**, **emitter** and **collector**. Normally the transistor will not let current flow between the emitter and collector but if a small current flows into the base, then the transistor will conduct.

The current flowing into the base turns the transistor on.

For many purposes the transistor is used with a **potential divider** arrangement. This is where two resistors are connected in series so that the total voltage available is divided across them.

To control a circuit, one of the resistors is replaced by a sensor such as a thermistor or a light dependent resistor (LDR).

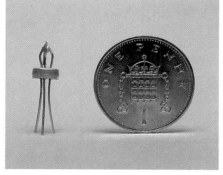

A transistor

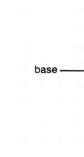

transistor symbol

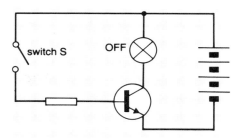

With switch S open no current can flow into the transistor. The transistor does not let current flow through the lamp.

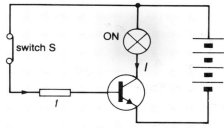

With the switch closed a small current flows into the base of the transistor. The transistor turns on and lets a large current flow through the lamp.

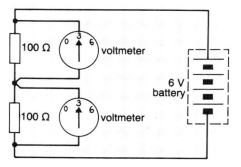

With equal resistances the total voltage is divided equally 3 V + 3 V = 6 V

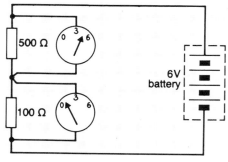

With unequal resistances more of the total voltage is across the larger resistor 5 V + 1 V = 6 V

Transistor-controlled circuits

Light-operated switch

A light can be automatically controlled using a transistor circuit. During the day time, when it is bright, the LDR has a lower resistance than the fixed resistor. The voltage across it is therefore relatively small so the current into the base is too small to make the transistor conduct. The light stays off.

As night approaches and it becomes dark, the resistance of the LDR increases. As a result the voltage across it increases causing a larger base current. When the base current is high enough, the transistor switches on and the bulb lights.

This circuit could be used to control the parking lights of a car. During the day the LDR would have a low resistance and so the lights would stay off. At night the LDR would have a high resistance and so the car's lights would be turned on.

Controlling a heater

In theory we could control a heater using a circuit similar to the light-operated switch but using a thermistor as a sensor.

Here the thermistor has a low resistance when it is hot. This means that it has a small voltage across it and so the base current is too small to turn the heater on. When the thermistor gets cold, its resistance rises increasing the voltage across it. This gives a bigger base current which turns the heater on.

Unfortunately, if the heater is to do any good it will need to carry a large current. (A single 1 kW bar of an electric fire carries about 4 A.) This large current would damage the transistor and so destroy our control circuit. To overcome this the transistor is used to control a **relay**. This is an electromagnetic switch which closes when a small current flows in its coil.

Now when the transistor turns on, a current flows in the coil of the relay. This closes the switch allowing a large current to flow through the heater. This heats the room.

Notice that this circuit is acting as a thermostat. As the heater heats the room the resistance of the thermistor gets less until the transistor turns the heater off again.

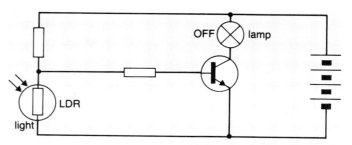

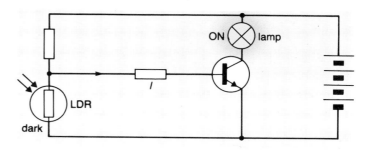

In the light the resistance of the LDR is low; the voltage across it is low; only a very, very small current flows into the transistor. The lamp turns off.

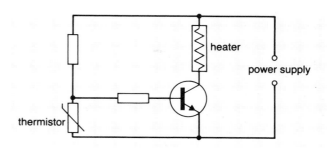

In darkness the resistance of the LDR is high; the voltage across it is high; a current flows into the transistor. The lamp turns on.

In this circuit there is a danger that the large current through the heater will destroy the transistor.

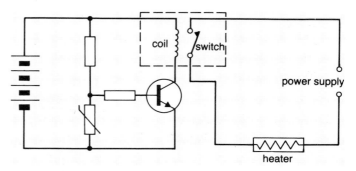

Using the transistor to operate a relay

Making decisions electronically

Circuits with switches in can be used for control as we have seen, but where there is more than one switch in a circuit we can think of the circuit as making decisions.

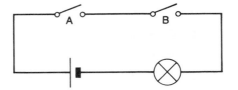

Bulb lights when switch A AND switch B are closed.

Bulb lights when switch A OR switch B (OR both) are closed.

Electronic control uses circuits like these but with electronic switches. These switches can be transistors (see page 168). They can also be packaged as **integrated circuits**. An integrated circuit is a piece of semiconducting material, for example silicon, which has many transistors built into it. This is then encased in plastic with small metal 'legs' or pins. The pins allow electrical connections to be made to the tiny circuits on the integrated circuit. The whole package is called a **chip**.

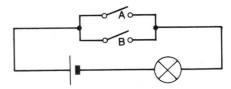

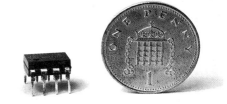

This integrated circuit 'chip' contains many transistor circuits.

Chips can be bought with circuits which act like switches connected in different ways. Each combination of switches is called a **gate**. This is because it acts like a gate, only letting information through when conditions are right. They are sometimes called **logic gates**.

Each gate has at least one **input** where voltages can be applied. We only need to consider two levels of input; a **high** input at about 6 V and a **low** input of about 0 V. Depending on the input, the gate then gives either a high or low voltage output. (It is convenient to call a high voltage '1' and a low voltage '0'.)

The table below shows how logic gates behave. The **truth tables** in the last column shows the output (1 or 0) for all combinations of inputs.

Type of gate	Symbol	What it does	Truth table
AND	A —, B — C	Gives a high output at C when inputs A AND B are high.	A B C / 0 0 0 / 0 1 0 / 1 0 0 / 1 1 1
OR	A —, B — C	Gives a high output at C when input A is high OR input B is high OR when both inputs are high.	A B C / 0 0 0 / 0 1 1 / 1 0 1 / 1 1 1
NOT	A — C	Only has one input. Gives a high output at C when the input is NOT high.	A C / 0 1 / 1 0
NAND	A —, B — C	Is the opposite of an AND gate; NAND = NOT AND. Gives a low output when A and B are high. Gives a high output for all other inputs.	A B C / 0 0 1 / 0 1 1 / 1 0 1 / 1 1 0
NOR	A —, B — C	Is the opposite of an OR gate; NOR = NOT OR. Gives a high output at C when neither A NOR B are high.	A B C / 0 0 1 / 0 1 0 / 1 0 0 / 1 1 0

Logic gates in control

Logic gate inputs can be connected to sensors. The gate will then only give an output when conditions from the sensors are 'correct'. This is very important for control. A simple example is shown opposite. It is an electronically operated lock which needs two keys to open it. Each key, when turned, gives a high input signal ('1'). If a high output signal is sent to the coil of the lock the door can be opened. The lock only opens when keys are turned in locks A AND B.

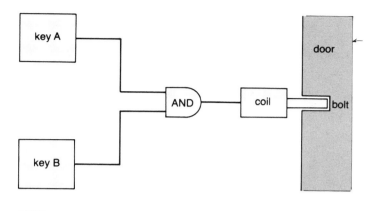

Now consider this; a bank manager has one key to the bank's safe. Each of the bank's two assistant managers also has a key. The manager wants to make sure that the safe is never opened unless he is there with at least one of his assistants. The circuit opposite could be used. The OR gate gives an output of '1' when key B OR C is turned. Now when the manager's key, A, is turned the AND gate gives an output of '1' to open the lock.

Some of the transistor circuits we have seen can be made using logic gates. Once again the sensor is used in a potential divider arrangement.

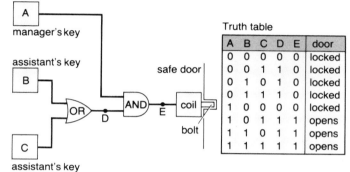

Truth table

A	B	C	D	E	door
0	0	0	0	0	locked
0	0	1	1	0	locked
0	1	0	1	0	locked
0	1	1	1	0	locked
1	0	0	0	0	locked
1	0	1	1	1	opens
1	1	0	1	1	opens
1	1	1	1	1	opens

Parking light circuit

Input A of the AND gate is always high because it is connected to the 6 V wire. When it is dark the resistance of the LDR is high and so input B is at a high level. The truth table shows us that when A AND B are high the output is high. This allows a small current to flow in the coil of the relay turning the light bulb on. When it is light the resistance of the LDR is low and so input B is low. The AND gate then gives a low output and the lamp turns off.

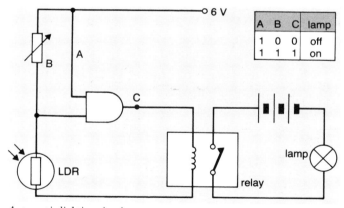

A	B	C	lamp
1	0	0	off
1	1	1	on

Automatic lighting circuit

Low temperature warning device

This device is designed to make a light-emitting diode (LED) light up when the temperature falls to 0 °C. When it is hot the thermistor has a low resistance. This means that input B is at a low voltage. The output of the AND gate is low and so the LED does not light. When it is cold the thermistor has a high resistance. This puts input B at a high level and so the gate gives a high output. The LED lights. The variable resistor can be adjusted to make sure that the LED comes on at 0 °C.

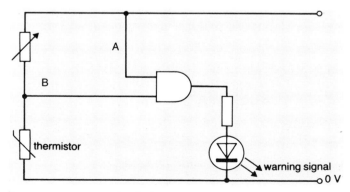

Low temperature warning device

Complex control circuits

Logic gates are ideally suited for use in complicated pieces of machinery where there are several things to be monitored and controlled. A good example is the automatic washing machine.

An automatic washing machine is very complicated but we can write some simple rules which must be followed by the machine if it is to work properly.

1   The machine is connected to a water supply which must be controlled.

2   The water has to be at the right temperature so a heater is used.

3   The machine must not fill with water if the door is open!

To build a suitable control system we need **sensors**, **actuators** and **logic gates**.

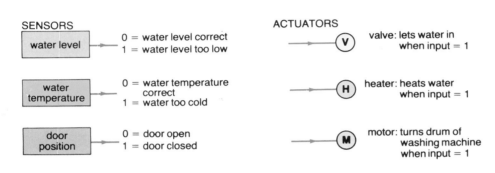

This washing machine is electronically controlled.

The circuit shown can be used to control the machine.

1   When the door is closed (output 1) **AND** the water level is low (output 1) the valve is opened and the washing machine fills with water. When the water level is correct the water level sensor gives an output of 0 so the valve closes and stops any more water from entering the machine.

2   When the door is closed (output 1) **AND** the water level is correct (output 0) **AND** the temperature is low (output 1) the heater is turned on. When the correct temperature is reached, the water temperature sensor gives an output of 0 so the heater turns off.

3   Finally, when the door is closed (output 1) **AND** the water level is correct (output 0) **AND** the water temperature is correct (output 0) the motor turns the drum of the washing machine.

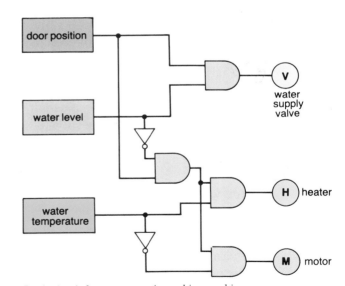

Logic circuit for an automatic washing machine

Questions

1   Name sensors that could be used
   a) to detect the water level in the machine
   b) to detect whether the door was open
   c) to detect whether the water was hot enough.

2   The water level sensor gives a '0' when the level is correct. What gate is used to turn this into a '1'? Why is this extra gate needed?

3   In practice, the logic gate chips cannot be directly connected to the heater circuit or to the motor. Why do you think this is?

Questions

1 Suggest suitable sensors for the following systems:
a) a greenhouse 'frost warning' device to warn a gardener when the air temperature is below 0°C
b) a warning device to tell a motorist when the petrol level is low
c) a warning device to sound an alarm when the door to a safe is opened
d) an alarm to wake a fisherman at dawn when the sun comes up.

2 A girl walking in the country accidentally touches an electric cattle fence with her hand. She immediately pulls her hand away. Explain what is happening using the following terms:
central nervous system, muscle, receptor in skin, motor neurone, sensory neurone, spinal cord.

3 The diagram below shows a simple bimetal thermostat controlling an electric heater. Explain how it keeps the temperature constant.

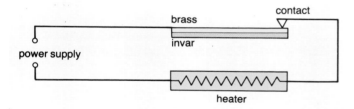

4 a) Explain how the body reacts when your internal body temperature starts to drop below its normal level.
b) Old people often suffer from hypothermia (very low internal body temperature) in cold weather. Suggest how hypothermia can be prevented.

5 The circuit below is a device which can sound a warning when a tank is full of rain water.

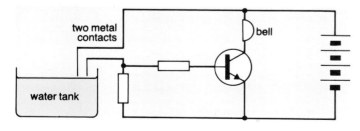

Explain how this works. (*Hint:* rain water is a good conductor of electricity.)

6 The diagram below shows some logic gates joined together.

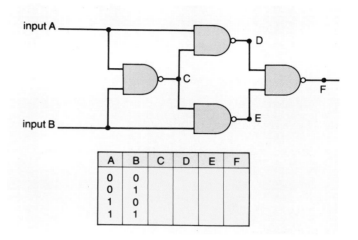

A	B	C	D	E	F
0	0				
0	1				
1	0				
1	1				

Copy and complete the truth table for the circuit. Describe the action of the circuit in words.

7 In many cars a carburettor is used to mix the right amount of petrol with air before it is burnt in the engine. The diagram below shows a simple carburettor.

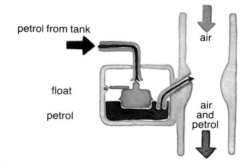

Simple carburettor

a) i) What happens to the float as more petrol enters the carburettor?
ii) How does this stop too much petrol getting in?
b) i) What happens to the float as petrol leaves the carburettor?
ii) Explain how this keeps the petrol level constant.
c) Where in the home would you find a control system which uses a float?

What is 'respiration'?
How much energy is there in our food?
Do all organisms need oxygen?
How can we use fermentation?

Scuba diver

Tree

Yoghurt (yogurt)

Fermenting wine

The photographs above all have something to do with living things. The person and the tree are large organisms and the wine and the yoghurt contain micro-organisms too small to see with the naked eye. These, and all other living things, need energy to stay alive. They get this from food where it is stored as chemical energy.

This chemical energy is released by a process called **respiration**. Respiration takes place inside the cells of plants and animals. Here energy is released from food in a chemical reaction. Usually this reaction needs oxygen. Respiration which uses oxygen is called **aerobic respiration**.

In Chapter 1 you will find a description of how **breathing** gets oxygen into your lungs and how this passes into the cells of your body. You will also find an explanation of **digestion** where the food you eat is broken down and how simple food substances get to the cells. This chapter explains how the cells release energy as you **respire**.

Questions

1 What is a ' micro-organism'?

2 What is 'respiration'?

3 What *two* substances do cells need for respiration?

4 What does 'aerobic' mean?

5 The photograph on this page shows a person wearing breathing apparatus.
  a) Give two other situations where breathing apparatus is needed.
  b) What is the *difference* between breathing and respiring and how are they connected?

Respiration: it's all about energy

In Chapter 1 we saw that carbohydrates and fats are high energy food substances. Digestion breaks these down into simpler molecules such as glucose. It is the energy holding these simple molecules together that is released during respiration.

We can summarize aerobic respiration in a chemical equation:

glucose + oxygen → carbon dioxide + water + ENERGY

$$C_6H_{12}O_6 + 6O_2 \rightarrow 6CO_2 + 6H_2O + 2880\,kJ$$

If you have studied the chapter on fuels you may notice that this form of respiration is very similar to burning a fuel. We can use this fact to measure the energy stored in different types of food.

Measuring the amount of energy stored in food

As the food, in this case sugar, burns it releases heat energy. This heat passes into the water causing the temperature to rise. The more energy in the food, the higher the temperature of the water will go. The table shows the results for sugar, a peanut and some dried bread.

table of results	sugar	peanut	dried bread
vol of water in test tube	50 ml	50 ml	50 ml
mass of water	50 g	50 g	50 g
mass of food	1 g	1 g	1 g
temp of water at start	21°C	21°C	21°C
temp of water at finish	71°C	91°C	51°C
temp rise	50°C	70°C	30°C

By using this formula it is possible to calculate the amount of energy (in joules) in the food).

energy = $\dfrac{\text{mass of water (in grams)} \times 4.2 \times \text{temperature rise (in °C)}}{\text{mass of food (in grams)}}$

For example:

energy in sugar = $\dfrac{50\,g \times 4.2\,J/°C \times 50\,°C}{1\,g}$

$$= 10\,500\,J \text{ or } 10.5\,kJ$$

Now calculate the energy value of the peanut and the dried bread.

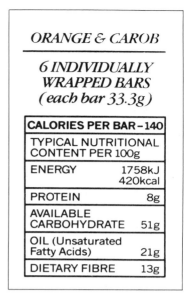

ORANGE & CAROB	
6 INDIVIDUALLY WRAPPED BARS (each bar 33.3g)	
CALORIES PER BAR – 140	
TYPICAL NUTRITIONAL CONTENT PER 100g	
ENERGY	1758kJ 420kcal
PROTEIN	8g
AVAILABLE CARBOHYDRATE	51g
OIL (Unsaturated Fatty Acids)	21g
DIETARY FIBRE	13g

Nutritional value tables include energy values.

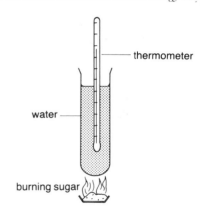

Energy from burning sugar

Activities

1 The experiment described gives us a rough idea of the energy value of foods. However, the number of joules calculated is *not* very accurate.
 a) Suggest why the result is not very accurate. Give at least three reasons.
 b) Design a better apparatus for this experiment. Draw a labelled diagram of your idea.

2 The energy value of food is often given on the packet.
Find the energy value in kilojoules (kJ) of the following foods:
 a) a chocolate bar ('Mars' or similar)
 b) baked beans
 c) tinned fruit in syrup
 d) breakfast cereal
Which one has the most energy in each gram? (You may need a calculator!)

Respiration in cells

The photograph opposite was taken using an electron microscope. It shows a cell enlarged 24000 times. The rod-like structures are called **mitochondria**. These are where respiration takes place.

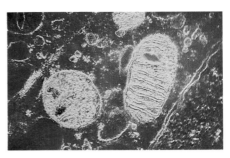

Mitochondria are found in the cytoplasm of all cells but the number varies. Muscle cells, for example, have lots of mitochondria because they need to release large amounts of energy quickly for movement.

The mitochondria in this cell are where respiration takes place.

Storing energy

The energy released during respiration is needed for many things. We need it for movement and to keep our body temperature steady (see chapter 8). As a result it is very important that our bodies should be able to store energy, as chemical energy, ready for use.

For long-term storage the body uses fat molecules but these cannot be broken down quickly. Cells must store energy for quick release when necessary. They do this using a chemical compound called ATP. (The letters stand for *a*denosine *tri*phosphate. Adenosine is a complicated molecule to which three phosphate groups are attached. A phosphate is made up of one phosphorus atom joined to four oxygen atoms.)

We can draw a diagram to represent ATP.

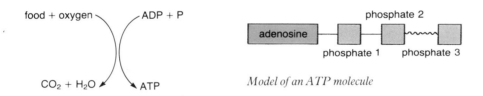

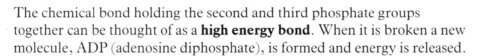

Model of an ATP molecule

The chemical bond holding the second and third phosphate groups together can be thought of as a **high energy bond**. When it is broken a new molecule, ADP (adenosine diphosphate), is formed and energy is released.

ATP is a short-term energy store in the cells which can release energy quickly when needed.

When the cell has excess energy, ADP molecules can be joined to phosphate groups to make ATP again.

Questions

1 What are 'mitochondria'?
2 Explain why muscle cells have more mitochondria than bone cells.
3 **a)** Explain why people get fat when they eat too many carbohydrates.
  b) Explain why strenuous exercise helps people to lose weight.
4 Why do muscle cells need a short-term energy store?
5 Describe what happens to ATP in the muscle cells of a rabbit when it suddenly sees a fox nearby.

Energy without oxygen (1): in animals

Most living cells in both plants and animals respire aerobically. That is, they use oxygen in the release of energy from food substances. However, sometimes an animal's breathing rate cannot get oxygen to the cells quickly enough. For example, if you are running a race you may create an oxygen shortage in your cells. (You may have heard sports commentators refer to this as an 'oxygen debt'.) Your muscles need to release more energy for movement but oxygen cannot get to the cells fast enough. So, where does the energy come from under these conditions?

ATP seems like the obvious answer but unfortunately ATP is only a short-term energy store and it is quickly used up. The body must therefore 'borrow' some energy from glucose in its cells. It does this by breaking the glucose down, without oxygen, into a substance called **lactic acid**. Lactic acid is a sort of halfway stage between glucose and its breakdown products, carbon dioxide and water. Energy is released and so you can continue running.

However, lactic acid builds up in the muscle cells causing muscle fatigue and eventually painful **cramp** – it all depends upon how fit you are!

After the exercise (or when cramp has forced you to stop) you usually gasp for air taking in lots of oxygen, your heart will also be beating faster to get more oxygen to the cells.

Athletes need to release energy quickly.

The lactic acid is slowly converted into carbon dioxide and water releasing more energy which is used to rebuild ATP molecules.

The process can be summarized as follows:

$$C_6H_{12}O_6 \rightarrow 2C_3H_6O_3 + 150\,kJ$$
$$\text{glucose} \qquad \text{lactic acid} \quad \text{energy}$$

Since the process brings about the release of energy from food without oxygen it is called **anaerobic respiration**.

Questions

1 What is the main difference between aerobic and anaerobic respiration?

2 Explain why a person creates an oxygen shortage (debt) during vigorous exercise.

3 Why is lactic acid called an 'intermediate' breakdown product of glucose?

4 What is cramp and how is it caused?

5 Explain how the body gets rid of excess lactic acid.

6 Footballers sometimes get cramp, particularly during a long hard game. When the trainer comes on to the field he can often be seen rubbing the footballer's legs vigorously. Suggest how this action helps overcome cramp.

After the race oxygen is needed for recovery!

Energy without oxygen (2): in plants

Plant cells, like animals cells, can produce energy by anaerobic respiration if necessary. This time however, the intermediate product is not lactic acid but an alcohol called ethanol. The following equation summarizes the process:

$$\text{glucose} \rightarrow \text{ethanol} + \text{carbon dioxide} + \text{energy}$$
$$C_6H_{12}O_6 \qquad 2C_2H_5OH \qquad 2CO_2 \qquad 210\,kJ$$

Germinating seeds and plant roots living in water-logged soil can respire without oxygen for a short time. However, the plant must return to aerobic respiration before the level of ethanol in the cells becomes too high, otherwise it will die.

Yeast, a microscopic fungus, can respire anaerobically or aerobically, depending on oxygen levels.

When little or no oxygen is present yeast breaks down glucose into ethanol and carbon dioxide with the release of energy; a process commonly known as **fermentation**. The yeast uses the energy to live.

Fermentation has been used for many hundreds of years in both brewing and baking. In brewing it is the alcohol that is used in beers, wines and spirits. In baking the carbon dioxide produced in fermentation makes bread dough rise.

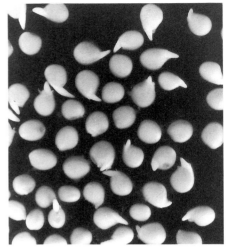

Germinating seeds can respire without oxygen.

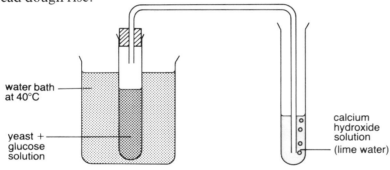

Investigating fermentation

Alcoholic fermentation in yeast

If some yeast is mixed with glucose solution and put in a warm place it will begin to ferment. A suitable apparatus for observing fermentation is shown in the diagram.

As the yeast breaks down the glucose to release energy a colourless gas is produced. This gas turns calcium hydroxide solution (limewater) milky.

What is this gas?

If the temperature of the water in the water bath is lowered by adding ice cubes, fewer bubbles of gas are produced. However, if the water temperature is raised to about 90 °C no bubbles are produced at all.

What effect do high temperatures have upon alcoholic fermentation in yeast?

Questions

1 A house plant needed watering once a week. When its owner went on holiday a 'kind' neighbour watered it once a day for two weeks. Explain why the plant died.

2 What is yeast? What does yeast produce when it respires anaerobically?

3 Yoghurt contains living bacteria. Bacteria are also capable of anaerobic respiration. If a sealed carton of yoghurt is left for a long time the lid bulges upwards. Suggest why this happens.

4 When yeast is used to make bread, the dough should be left in a warm place so that the yeast can produce carbon dioxide to make it rise. Explain why the dough would not rise if it were put straight into a hot oven.

5 Why is the amount of energy produced during fermentation much less than that produced by aerobic respiration?

Carbohydrates: the energy foods

Carbohydrates are food substances which can be used in the body to release energy. All carbohydrates contain carbon, hydrogen and oxygen atoms. Examples are the sugars, glucose ($C_6H_{12}O_6$) and maltose ($C_{12}H_{22}O_{11}$), and starch. The starch molecule has hundreds of sugar molecules joined together.

Sugars

Glucose is one of the simplest sugars. Its formula is $C_6H_{12}O_6$ and it has the ring shape shown in the diagram. There are other sugar molecules with this formula but their atoms are arranged differently. Molecules with the same formula but different structures are called **isomers**. The diagrams show two isomers of glucose; fructose and galactose.

Sugars provide energy . . . but can cause tooth decay.

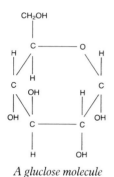

A gluclose molecule

Fructose

Galactose

Glucose, fructose and galactose are called single sugars or **monosaccharides**. They all have slightly different properties but they all taste sweet.

Two single sugar molecules can be linked together to form a double sugar or **disaccharide**. Maltose is a double sugar made when two glucose molecules join up with the loss of a water molecule. Reactions like this in which water is removed are called **condensation reactions**.

$$glucose + glucose \rightarrow maltose + water$$
$$C_6H_{12}O_6 + C_6H_{12}O_6 \rightarrow C_{12}H_{22}O_{11} + H_2O$$

Maltose, sucrose and lactose are all disaccharides. These also taste sweet.

Sugars in food

All sugars taste sweet and because people like the taste we use a lot in our food and drink.

Sugars are also soluble in water. This enables the dissolved sugar to be carried in the transport systems of animals and plants.

Unfortunately sugar on our teeth encourages plaque and so causes tooth decay (see Chapter 1).

Many foods now carry lists of what they contain. If you look at the labels you may be surprised at some of the foods that have sugar added to them; these baked beans for example.

Questions

1 Look at the diagrams of the glucose and fructose molecules.
   a) List *two* similarities between them.
   b) List *one* difference between them.

2 Why are glucose and fructose called 'carbohydrates'?

3 Why are glucose and galactose called 'isomers'?

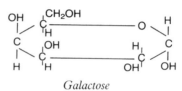

INGREDIENTS		
Beans, Tomato Purée, Water, Sugar (2.2%), Salt (0.5%), Modified Starch, Onion Powder, Spices.		

NUTRITION		
A serving = approx. ⅓ of the can.		
AVERAGE COMPOSITION	PER 140g (5oz) serving	PER 100g 3½oz
Energy	353kJ/84kcal	252kJ/60kcal
Fat	0.6g	0.4g
Protein	7.0g	5.0g
Available Carbohydrate	13.3g	9.5g
Fibre	10.2g	7.3g
Added Salt	0.7g	0.5g
Added Sugars	3.0g	2.2g

Even baked beans contain sugar!

Starch: a big carbohydrate molecule

We have seen how two glucose molecules can be joined to give a maltose molecule. The two monosaccharides join to give a disaccharide. In fact many single sugar molecules can join together in condensation reactions. The larger sugar molecules formed are called **polysaccharides**.

Starch is a polysaccharide with several hundred condensed glucose molecules linked together in a chain. Its formula is written as $(C_6H_{10}O_5)_n$ where n is the number of glucose 'building blocks' linked together.

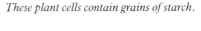

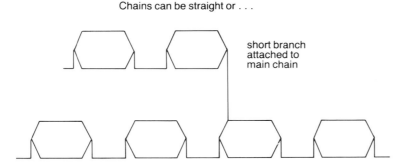

These plant cells contain grains of starch.

glucose molecules

Chains can be straight or . . .

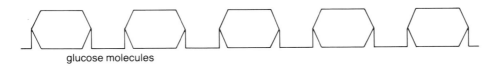

short branch attached to main chain

. . . chains can be branched.

Starch does not dissolve in water. Because it is insoluble it can be used as a way of storing glucose. In plants the starch molecule chains are folded and bundled together to form starch grains like the ones shown in the photograph at the top of the page. Starch grains are found in large numbers in rice, wheat and other cereal crops, all of which are grown for food. The starch stored in animals is a little different to that found in plant cells. Animal starch is called **glycogen**.

Rice has a high starch content.

Questions

1　What is the difference between a monosaccharide, disaccharide and a polysaccharide?

2　What are the building blocks of starch molecules?

3　**a)** Give two differences between starch and glucose.
　b) Give one similarity between starch and glucose.

4　Write the formula for a starch molecule that is made of 500 glucose molecules.

5　Starch is found in plants and animals as a food store.
　a) What makes starch molecules ideal as a food store?
　b) Name two foods that we eat that contain a lot of starch.
　c) Why do we eat starchy foods?
　d) What happens if we eat too many starchy foods?

Carbohydrates: building them up and breaking them down

The removal of a water molecule from two monosaccharide molecules produces a disaccharide molecule. If a water molecule is added to a disaccharide two monosaccharides will be produced. In other words, **condensation is reversible**.

The process of breaking up molecules with the addition of water is called **hydrolysis**. Hydrolysis is the direct opposite of condensation.

The series of diagrams opposite show how carbohydrates are both built up and broken down.

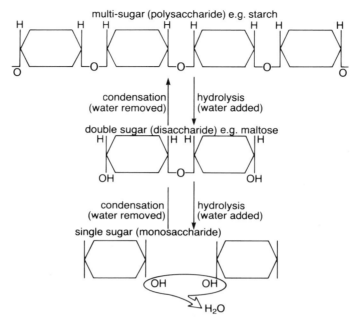

Condensation and hydrolysis in sugars

Hydrolysis of starch molecules

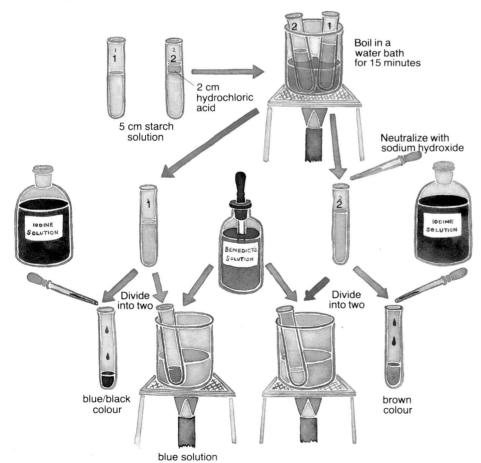

Questions

1 What do you suppose the acid has done to the long starch molecule chains?

2 What was the point of adding sodium hydroxide to test tube 2 before testing with iodine and Benedict's reagent?

3 In this experiment what proves that it is the hydrochloric acid that causes the change in the starch molecules?

More about fermentation

This is fermentation on a very large scale.

This bakery produces about 500 000 loaves of bread a week.

This brewery produces over 54 000 000 litres of lager a week.

Bakers add yeast to dough to make it rise. The yeast feeds on the sugars in the dough and as it respires it produces carbon dioxide. The gas forms bubbles in the dough because it cannot escape. These are the holes that you can see in baked bread. During baking the yeast is killed and fermentation stops.

In wine and beer making it is the other by-product of fermentation, alcohol, which is required. Wine is made as yeast feeds on the sugars in fruit. The type of wine produced depends upon the type of fruit used. Most commercial wine is made from grapes.

Beer and lagers are made from barley. Barley seeds are allowed to germinate until all the seed food stores are converted into sugar. The sugar is in fact maltose (often referred to as 'malt'). Boiling water is then added which stops germination and dissolves the sugar. Yeast is mixed into this sugar solution and the mixture is allowed to ferment. Flavourings are added to beers to produce different varieties. For example, hops give bitter beer its distinctive taste.

Spirits such as whisky, gin and brandy are made from wine or the sugar from germinating barley. Brandy is made by distilling wine, whisky comes from the distillation of fermented maltose solution. Since yeast cannot live in high concentrations of alcohol, wines and beers have a limited strength (maximum 20 per cent alcohol). Distillation provides a means of extracting nearly all of the alcohol produced during fermentation. Spirits are much stronger than wine.

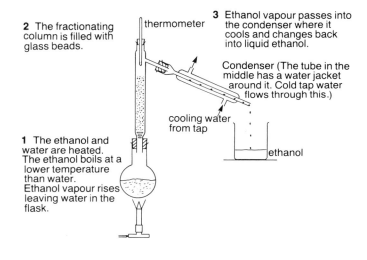

2 The fractionating column is filled with glass beads.

thermometer

3 Ethanol vapour passes into the condenser where it cools and changes back into liquid ethanol.

Condenser (The tube in the middle has a water jacket around it. Cold tap water flows through this.)

cooling water from tap

1 The ethanol and water are heated. The ethanol boils at a lower temperature than water. Ethanol vapour rises leaving water in the flask.

ethanol

Distillation of ethanol

Questions

1 What do you suppose happens to the alcohol produced by the yeast during the making of bread?

2 When making some wines extra sugar must be added to the fruit and yeast mixture. Why should this be necessary?

3 Scottish malt whisky is world famous. How do you think it got its name?

4 Describe how brandy is made.

Alcohols

The proper chemical name for alcohol produced by anaerobic respiration in yeast is ethanol, C_2H_5OH. Ethanol is an important member of a group of chemicals which form a **homologous series**. This means that all of the members of the group have a similar shape of molecule and have the same general formula. Alcohols like ethanol have the general formula $C_nH_{2n+1}OH$. Notice the OH group at the end. When n=1 the alcohol is CH_3OH, methanol, when n=2 the alcohol is C_2H_5OH, ethanol, and so on.

The table below gives you more information about these and other members of the alcohol family.

Name	Molecular formula	Structural formula	Freezing point	Boiling point	Solubility in water
Methanol	CH_3OH	H | H—C—O—H | H	−97°C	64.1°C	∞
Ethanol	C_2H_5OH	H H | | H—C—C—O—H | | H H	−117°C	78.5°C	∞
Propanol	C_3H_7OH	H H H | | | H—C—C—C—O—H | | | H H H	−127°C	97.4°C	∞
Butanol	C_4H_9OH	H H H H | | | | H—C—C—C—C—O—H | | | | H H H H	−90°C	117.4°C	79 g/litre

The boiling point is the temperature at which liquid turns into a gas. Look what happens to the boiling point as the molecules get larger; it rises. This is because small molecules have weaker forces holding them together than larger molecules. So, when they are heated it is the smaller ones that fly apart first, forming a gas.

The solubility of alcohols is also linked to the size of their molecules. Those with large molecules are almost insoluble in water. Those with smaller molecules dissolve easily.

Questions

1 Using the table answer the following questions.
   a) Write down the molecular formula of the alcohol with 5 carbon atoms (n=5).
   b) Give the structural formula of the alcohol with 7 carbon atoms (n=7).
   c) Estimate the boiling point of the alcohol with 6 carbon atoms (n=6).
   d) Describe what happens to the freezing point as the alcohol molecule gets bigger.
   e) Give *two* reasons why ethanol could be used as antifreeze in a car radiator. (*Hint:* water freezes at 0°C.)

Antifreeze

Drinking alcohol

A large number of people drink alcohol at sometime in their lives. It may be in wine, beer or spirits drunk at social gatherings in homes and public houses. Some people take alcoholic drinks every day. Others just drink alcohol on special occasions such as weddings, birthdays or at Christmas.

Alcohol is absorbed very quickly through the walls of the digestive system into the blood. The amount of alcohol absorbed depends upon how much is drunk, whether food is eaten at the time of drinking and even upon what height, weight and sex the person is. It is a fact that women are affected more by alcohol because their bodies do not contain as much water as those of men and so the alcohol is not diluted as much. The liver removes alcohol from the blood during normal circulation but it can take a long time before it is all removed.

Not all drinks have the same alcohol content. The photograph shows a range of drinks each one containing the same amount of alcohol – this quantity is often referred to as **one unit** of alcohol.

Doctors advise men not to drink more than 20 units of alcohol in one week. Women should not drink more than 13 units per week. If people keep within these limits experts believe that there is little risk of long term damage to health. The table gives you more information about the number of units of alcohol in popular drinks.

How many 'standard drinks' in your drink?		
		standard drinks
beers and lagers		
ordinary strength beer or lager	$\frac{1}{2}$ pint	1
	1 pint	2
	1 can	$1\frac{1}{2}$
Strong ale or lager	$\frac{1}{2}$ pint	2
	1 pint	4
	1 can	3
Extra strength beer or lager	$\frac{1}{2}$ pint	$2\frac{1}{2}$
	1 pint	5
	1 can	4
ciders		
average cider	1 pint	3
strong cider	1 pint	4
spirits		
1 standard single measure in most of England and Wales ($\frac{1}{6}$ gill)		1
1 standard single measure in Northern Ireland ($\frac{1}{4}$ gill)		$1\frac{1}{2}$
$\frac{1}{3}$ gill measure served in some parts of Scotland		2
1 bottle		30
table wine		
(including cider wine and barley wine)	1 standard glass	1
	1 bottle (700 ml)	7

Activities

Design a questionnaire and use it to gather information about the drinking habits of your friends and relatives.

Here are some questions to include.

'Do you drink alcoholic drinks never / occasionally / regularly?'

'Do you drink alcoholic drinks most days / every week / every month / rarely?'

'Do you normally drink beer / wine / spirits / a mixture?'

'In a **normal** week how much beer, wine and spirits do you drink?'

After you've finished gathering data, use your results to answer these questions.

1 What fraction of your friends and relatives drink alcohol at least once a week?

2 What fraction of your friends and relatives never drink alcohol (teetotal)?

3 Are any of your friends or relatives over the danger levels of 20 units per week for a man or 13 units per week for a woman?

Alcohol the drug

A small amount of alcohol makes people feel happy and relaxed but alcohol is really a **depressant** drug. It depresses or reduces brain activity and thereby affects judgement, self control and the time taken to react to a stimulus.

The level of alcohol in the blood is measured in milligrams per 100 millilitres of blood (mg/100 ml). Drinking a single unit of alcohol raises the concentration of alcohol in the blood by about 15 mg/100 ml.

The legal driving limit for blood alcohol level is 80 mg/100 ml but even when at this level a driver is three times more likely to have an accident than if they had not drunk at all. The only really safe limit when driving is no alcohol at all! Drunken driving is a major cause of road accidents. One third of drivers killed in such accidents have blood alcohol levels well in excess of the legal limit.

Drunken driving is a major cause of road accidents. One third of drivers killed in accidents have blood alcohol levels above the legal limit.

Questions

1 **a)** How many pints of beer would a person need to drink to be over the legal limit?
   b) Why would a small person need to drink less alcohol than a large person to be over the limit?
2 **a)** How many glasses of wine would a person need to drink to be over the legal limit?
   b) Explain why *two* glasses of wine at lunchtime and *two* pints of beer on the way home from work could put a driver over the legal limit.

Like many other drugs alcohol can be **addictive**. However, unlike other addictive drugs it is available in supermarkets, off-licences and pubs. Therefore people need to be careful about how much alcohol they drink otherwise they may become dependent; they become alcoholics. Alcoholics, like other drug addicts, run a great risk of serious damage to their health. It is a sad reflection on our society that more and more young people are becoming hooked on alcohol.

Consider the following facts:
* too much alcohol present in the blood over a number of years can lead to liver damage. **Hepatitis** (liver inflammation) and **cirrhosis** (scarring of liver tissue) are common amongst heavy drinkers and can cause death.
* some people often go without food and have a drink instead. These people are likely to develop stomach ulcers and other problems in the digestive system.
* a high blood alcohol level causes a rise in blood pressure. You will have already read in Chapter 1 that high blood pressure can be a contributory factor in coronary heart disease.

It is in everyone's interest to either avoid alcohol or to drink within sensible limits.

Questions

1 What is an 'alcoholic'?

2 How may alcohol damage the body of an alcoholic?

3 People with high-pressure jobs often drink alcohol at lunchtime without eating food. Why is this dangerous?

4 Explain why it is illegal for a shopkeeper or a publican to sell alcohol to a person under 18 years of age.

5 Suggest ways of reducing the number of school students who drink alcohol.

Enzymes: helping to release energy

Each atom in a molecule is held to neighbouring atoms by chemical bonds. Breaking these bonds can release some energy. For example, you already know that energy is released from glucose molecules when they break down into carbon dioxide and water. But what causes glucose to break down and give up its energy?

Glucose breakdown cannot happen unless a certain amount of energy is put in first. This is called **activation energy**. However, as the carbon dioxide and water are made, much more energy is released than was needed to start the reaction. We can think of this as an energy 'profit'. This is the energy produced in respiration. It can be used for maintaining body processes and building new body tissues and also to activate the breakdown of more glucose molecules!

The breakdown of glucose is rather like a rock being pushed over a cliff; energy must be put in to get the rock moving. Once over the edge, the rock falls, releasing more energy than was used to push it in the first place.

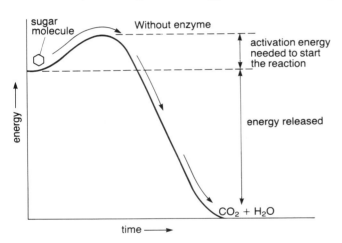

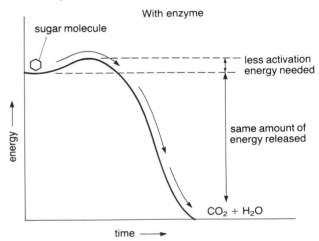

Enzymes reduce the activation energy 'hump'

You have seen that sugar releases energy when it is heated by a flame. The high temperature provides the energy required to activate some sugar molecules into breakdown. The energy released causes more heat and yet more energy release, and so on until the sugar is all burnt.

But how does respiration take place at the relatively low temperatures found in the cells of living organisms?

The answer is to use a **catalyst**. A catalyst is a substance which speeds up a chemical reaction but remains unchanged at the end. Biological catalysts are called **enzymes**. They lower the amount of activation energy required for chemical reactions such as respiration to proceed.

Questions

1 What holds the atoms together in a molecule?

2 What is 'activation energy'?

3 A bowl of sugar on a kitchen table does not suddenly burst into flames. Sugar thrown on to a fire burns brightly and releases lots of energy.
Explain these two facts.

4 What is a catalyst? Why are catalysts (enzymes) important for respiration in cells?

Enzymes: what affects them

The activity of enzymes is affected by changes in acidity (pH) and temperature. Each enzyme works best at a particular level of pH and temperature. Its activity reduces above or below that point.

Pepsin is an enzyme found in the stomach. It helps in the digestion of protein and works most effectively at pH2 (strong acid). Trypsin is found in the duodenum and only functions in alkaline conditions (pH8).

Enzymes are actually proteins and they are affected by **heat**. Heat changes the structure of proteins. You can see this by watching what happens to the white of an egg when it is poached or fried. Our normal body temperature is 37°C and, not surprisingly, the enzymes in our cells work best at this temperature. Most enzymes cannot tolerate temperatures higher than 45°C.

Usually enzymes are **specific**. This means that they will only catalyse one kind of reaction. Most enzymes work on one particular molecule, however some digestive enzymes are able to act on a range of closely related molecules. Lipase for example will break down a number of types of fat during digestion.

The effect of temperature on enzyme activity

If you have read Chapter 1 you will remember that iodine turns from brown to blue/black when it is added to starch.

If some amylase (the enzyme in saliva) is added to a mixture of iodine and starch solution this blue/black colour will slowly disappear. This is because the amylase has hydrolysed (broken down) the starch into sugar.

By noting the time taken for the blue/black colour to disappear we are able to work out how fast the amylase breaks down the starch. This is called the rate of reaction.

When this experiment is repeated at different temperature the rate of the reaction changes. The graph shows you this change.

We can easily see the effect of heat on the protein in an egg.

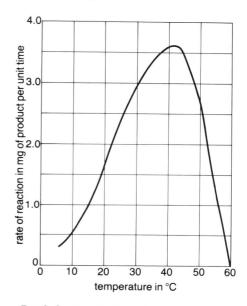

Graph showing the effect of temperature on enzyme activity

Questions

(Questions 1 and 2 refer to the graph.)

1 **a)** What is the rate of reaction
at   i) 10°C   ii) 20°C   iii) 30°C?
b) What happens to the rate of reaction every 10°C?

2 **a)** What is the temperature that gives the highest rate of reaction?
b) What happens to the rate of reaction above this temperature?
c) What do you suppose happens to the enzyme at around 60°C? (*Hint*: remember the egg!)

3 Amylase is one of the enzymes in your body. Your normal body temperature is about 37°C. When you get a disease your body temperature usually rises. Why do you suppose doctors take your temperature regularly when you get a bad infection of a disease?

4 **a)** Give two ways in which a mammal like an elephant can stop its body temperature from getting too high.
b) Why do you suppose 'cold blooded' animals like lizards need to warm themselves up by sunbathing before they can move quickly?

Enzymes: how do they work

We have seen that enzymes are very important in helping chemical reactions to occur in living organisms. They speed up reactions but they are not broken down or changed; so how do they work?

Scientists believe they have found the answer. The clue is that enzymes are very **specific**. This means that they only work on one or two different types of molecule. The molecule which is to be broken down is called the **substrate**.

Research shows that enzymes work by firstly attaching themselves to substrate molecules.

The fit of the enzyme and substrate must be exact otherwise the enzyme will not do its job. This is often referred to as the lock and key mechanism where the substrate 'key' must exactly fit the enzyme 'lock'.

The place on the enzyme molecule where the substrate fits is called an **active site**. Substrate molecules stay attached to enzymes at active sites until they are activated into forming the product molecules of the reaction. The whole process is extremely quick if conditions like temperature and pH are right.

enzyme + substrate → enzyme–substrate → enzyme + product molecules
complex

Notice that once the enzyme has done its job it is free to go on and catalyse the breakdown of more substrate molecules. This, together with the speed at which they work is the reason why cells can function perfectly well with only a tiny amount of enzyme. (It is also the reason why you need only use small amounts of enzyme solution in enzyme experiments!)

Looking to see how enzymes speed up chemical reactions

Hydrogen peroxide (H_2O_2) breaks down into water and oxygen very slowly under normal conditions. You can see the oxygen being released as tiny bubbles if you put some in a beaker.

However, if a small piece of fresh liver is added to the hydrogen peroxide see what happens.

The breakdown of the hydrogen peroxide speeds up rapidly. How could you prove that the gas given off was oxygen?

Questions

1 The formula shown below shows the breakdown of hydrogen peroxide.

$$2H_2O_2 \rightarrow 2H_2O + O_2$$

a) Name the two products of this reaction.
b) A student finds that a small piece of fresh liver added to some hydrogen peroxide in a test tube speeds up the release of oxygen. Design an experiment to find out whether the enzyme is sensitive to changes in pH.

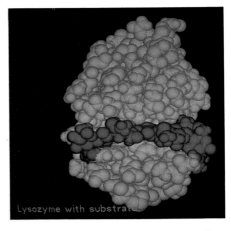

A model of the enzyme lysozyme showing the active site. Lysozyme is an enzyme found in tears. It kills bacteria by dissolving them! (The active site is in red.)

substrate molecule

1 *Substrate 'key' moves into enzyme 'lock'*

enzyme molecule

2 *Substrate fits enzyme forming an enzyme-substrate complex*

product molecules

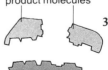

3 *Chemical bonds are broken and product molecules are released – the enzyme remains unchanged*

Questions

1 a) What is respiration?

b) The equation below represents one form of respiration.

$$C_6H_{12}O_6 + 6O_2 \rightarrow 6CO_2 + 6H_2O + ?$$
(Glucose)

i) Is this aerobic or anaerobic respiration?
ii) How do you know?
iii) Name the *two* products shown in the equation.
iv) What product does ? represent?

2 'In 1968 the Olympic games were held in Mexico City. This caused problems for long distance runners because at 2300 m above sea level the atmosphere is thin and contains less oxygen. Many athletes developed severe muscle cramps. Others had to be given pure oxygen to breathe after they collapsed.'

a) Why does the air in Mexico City 'contain less oxygen' than the air near sea level?

b) Why did the lack of oxygen cause runners to develop muscle cramps? (Your answer should include the terms *oxygen debt* and *lactic acid*.)

c) Suggest how breathing pure oxygen instead of air helps the athletes recover.

3 The label shown below is from a tin of pasta shapes.

CONSUMER CARE

NUTRITION INFORMATION 100 GRAMS OF THIS PRODUCT TYPICALLY PROVIDES	
12.0 grams of Protein	HIGH
75.0 grams of Carbohydrate	HIGH
2.0 grams of Fat	LOW
Energy value	·1480 kJ
(Calories)	(347 kcal)

a) Name *two* listed food substances which can provide energy.

b) Some of the carbohydrate is starch. Describe the structure of a starch molecule.

c) Why is starch useful in the body?

4 a) Describe the effects of alcohol on a person's reactions and general behaviour. Explain how these effects can make accidents more likely at work and in the street.

b) Alcohol is an addictive drug.
i) What is meant by 'addictive'?
ii) What laws do we have in our society to control drinking?

5 The diagram below shows some home made wine fermenting.

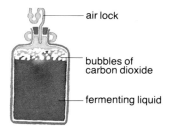

The liquid contains fruit juice, sugar and yeast.

a) What is fermentation?

b) i) One of the products of fermentation is a colourless gas which bubbles out of the air lock. Name the gas.
ii) The other product is a colourless liquid which, when drunk, affects the nervous system. Name the liquid.

c) Most home wine makers know that fermentation is faster when the glass jar is kept in a warm cupboard. However, an impatient wine maker decides to heat up the fermenting liquid until it boils. Why doesn't he get his wine?

6 This label is from a bottle of whisky.

a) Describe how the whisky was made.

b) The bottle contains 1000 ml (1 litre) of whisky. 43 per cent is alcohol.
i) Calculate the volume of alcohol in the bottle.
ii) 10 ml of alcohol has a mass of about 7 g. Estimate the mass of the alcohol in the bottle.
iii) The average body contains about 5000 ml of blood. The legal limit for driving is 80 mg of alcohol in 100 ml of blood.
Show that someone who drinks half a bottle of whisky is way over the limit.

What is a detergent?
How is soap made?
How do soaps and detergents clean things?
Does soap work in all types of water?
Can detergents damage our environment?

Detergents are chemicals which, when dissolved in water, can remove dirt and grease from cloth, metal, ceramics and, of course, human skin.

Soap is a form of detergent made from animal fats or plant oils. Other detergents can be made from chemicals extracted from crude oil (see page 80). These are often called soapless detergents.

A quick look around the house shows just how important detergents are in our lives. Apart from bars of soap you are likely to find shampoo, washing-up liquid, and washing powders. The manufacture of detergents is big business. The world uses over 15 million tonnes each year. It is hard to believe that less than 200 years ago ordinary people did not use soap. Only the rich could afford to buy soap and the Government taxed it as a luxury!

No one knows exactly how the method for making soap was discovered. It may have been by accident. We do know that it was used over 2000 years ago by the Phoenicians. They made it from wood ashes and fat from goats. We can guess that their soap was fairly soft and very smelly. This did not really matter because it was used to clean the grease from wool and cloth before it was dyed. This is still an important use of detergents.

Activities

1 Look in your home for chemicals used for cleaning. Record your results in a table using the headings below.

Trade name	What does it look like?	What is it used for?	Is it soap or detergent?

2 The painting shown on this page was used as part of an advertisement for soap. Cut out modern soap advertisements from magazines and make a poster. How do advertisers try to get us to buy their product?

3 Describe a 'soap powder' commercial shown on television. Does it use scientific language to sell the product? Does it compare washing powders in a scientific way?

4 Design a scientific experiment to compare the cleaning power of two washing powders.

Making soap

The first soap makers probably used wood ash from their fires mixed with animal fat. The mixture was boiled for a long time. More ashes were added as the liquid evaporated. Eventually a solid soap formed which could be used for washing cloth.

The early soap makers did not understand the complicated chemical reactions which produced the soap but the method used today still includes animal fats or plant oils. Strong alkali solutions have replaced the wood ash!

Making soap in the laboratory

Castor oil is an oil from a plant. When it is boiled with a strong alkali solution a reaction takes place. The reaction is called **saponification** because soap is produced.

DANGER

Be very careful. Sodium hydroxide is very corrosive. It can cause serious burns. Safety glasses must be worn.

10 cm³ concentrated sodium hydroxide solution

2 cm³ castor oil

Put 2 cm³ of castor oil into a clean beaker. Add 10 cm³ of concentrated sodium hydroxide solution.

stir with a glass rod

small flame for gentle boiling

Warm the mixture gently over a low flame until it just starts to boil. Stir it all the time using a glass rod. Keep it boiling gently for five minutes. The mixture may boil over so watch carefully and be ready to remove the heat source.

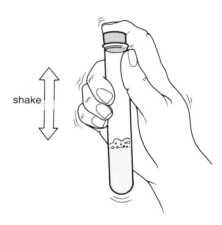

shake!

The solid trapped in the filter paper should be soap. Test it by putting a small piece in a test tube half filled with distilled water. Put a bung in the test tube and shake it vigorously. If it foams you have been successful! Your soap will still be alkaline so do not use it to wash with.

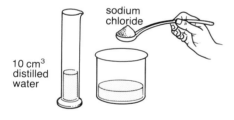

sodium chloride

10 cm³ distilled water

Remove the beaker from the heat. Add 10 cm³ of water and two heaped measures of sodium chloride (common salt).

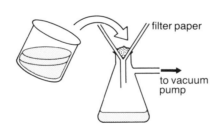

filter paper

to vacuum pump

Heat the mixture again stirring it all the time. When it boils pour it into a filter funnel. (Filtering will be quicker if a pump is used.)

The chemistry of saponification

The fats and oils used for making soap contain chemicals called **glycerides**. The glyceride molecule is complicated. It is made up of a small glycerol molecule with fatty acids joined to it. The fatty acids have long chains of carbon atoms in their structure.

When oil or fat is heated with an alkali such as sodium hydroxide solution, the long, fatty acid parts break away from the glycerol. These react with the sodium hydroxide to make a complicated sodium salt which we call soap. (In industry the glycerol is separated from the soap and used in the manufacture of cosmetics.)

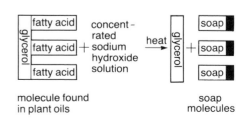

molecule found in plant oils

soap molecules

triglyceride + sodium → glycerol + 'soap'
hydroxide

Saponification

Making soap in industry

The method for making soap described on page 191 is used by soap makers for preparing small **batches** of special or high quality soaps. However, heating the oil and alkali solution in open pans takes a long time. The mixture then has to stand for several days to let the soap separate out. This makes the soap expensive. Most of the soap used in our homes is made by a **continuous process**.

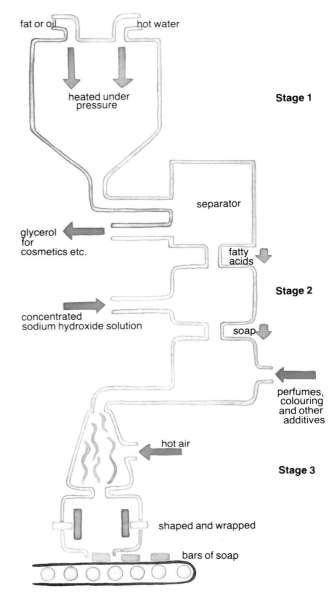

Making soap in industry

Stage 1
Molten animal fat or plant oil is pumped into a tall reaction column. Here it reacts with hot water under pressure. The reaction gives glycerol and fatty acids (see page 194).

The glycerol is separated from the fatty acids and used to make cosmetics. The fatty acids are used to make the soap.

Stage 2
The fatty acids are purified and then pumped into a reaction vessel with sodium hydroxide solution. These react to give soap. Because fatty acids are used there are no impurities to remove. Some coconut or olive oil is usually added at this stage to make sure that all the sodium hydroxide has reacted. Too much alkali in the soap could make delicate skin sore.

Questions

1 What raw materials are used for making soap?

2 What is a **batch** process? Suggest why special soaps are made in this way.

3 What two types of substance react to make soap in the continuous process? How are they mixed?

4 Explain why soap from the continuous process is cheaper than soap from the batch method.

5 Give three ways in which soap makers try to make their soaps more attractive to customers.

Stage 3
Things such as perfumes and colourings are added to the soap to prepare it for sale. It is then dried and pressed into shape before wrapping.

This continuous process takes just a few hours to produce the finished bar of soap. The batch process can take up to ten days!

How do soaps and detergents clean?

Soaps and detergents are very good at cleaning dirt and grease from things like cloth and skin. They do this by three actions:

1 They help water to **wet** things.

2 Their molecules can lift dirt from the surface.

3 Their molecules help to keep the dirt suspended in the washing water so that it does not stick to the surface again.

The wetting action of detergents

Water molecules attract each other. We can see the effect of this by carefully filling a glass with water until the water is just over the rim. The molecules attracting each other form a bulging surface which is a bit like an elastic skin. This effect is called **surface tension**.

The diagram shows some water on a piece of glass covered with petroleum jelly. The forces between the water molecules hold them into one large drop. They do not stick to the grease molecules and so the glass does not get wet.

Detergent molecules reduce the surface tension of water.

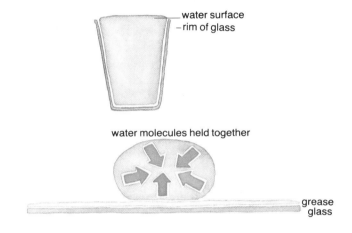

water surface
– rim of glass

water molecules held together

grease
glass

Activities

1 Fill a glass with tap water.
Cut a piece of filter paper or kitchen paper which will easily fit inside the glass.

Put a sewing needle on top of the paper and then gently lower the paper until it floats on the water. Leave the glass undisturbed.

After a few minutes the paper will sink leaving the steel needle supported by surface tension.

Pour one or two drops of detergent (washing-up liquid) down the side of the glass.

The needle sinks almost immediately. This is because the detergent has reduced the surface tension of the water.

2 Take a piece of dry, woven cloth. The denim used to make jeans is ideal.

Use a teat pipette to drop water on to the cloth.

A large drop should form. Draw its shape.

Add one drop of detergent to the water. See what happens.

The drop suddenly spreads out and soaks into the cloth.

water

sewing
needle

filter paper
or kitchen paper

needle held
on surface

soggy paper

detergent

needle sinks

To understand why soaps and other detergents are good for washing dirty clothes and greasy plates we need to know something about the structure of their molecules.

Chemistry of a soap molecule

The fats and oils used to make soap contain **organic compounds** (see page 81). These have about 50 carbon atoms in their long molecules. When they are heated with a strong alkali they react to form a compound called **glycerol** and a complicated sodium salt. One type of soap is made from the oil from a palm tree. A word equation for the reaction is given below.

palm oil + sodium hydroxide →

glycerol + sodium palmitate

Glycerol is an alcohol and sodium palmitate is a soap detergent. The diagram shows a model of sodium palmitate.

We can think of the soap molecule as having two parts; the end with the sodium and a long chain of carbon and hydrogen atoms.

The sodium end dissolves in water because it is **ionic**. The sodium ion is attracted to a negatively charged oxygen ion at the end of the long chain part of the molecule. This electrical attraction bonds the sodium to the rest of the soap molecule.

Ionic substances dissolve in water because the water molecule has a positive end and a negative end. In water the forces of attraction pull the ionic compounds apart and make them dissolve.

The sodium end of a soap molecule is sometimes called **hydrophilic**. This means 'water-loving'.

The carbon and hydrogen atoms in the long 'tail' of the soap molecule are not held together by ionic bonds. As a result it behaves a bit like paraffin. It will not dissolve in water but it will attach itself to grease. This part of the molecule is 'water-hating' or **hydrophobic**.

Soaps clean because the two ends of their molecules behave very differently. In diagrams a detergent molecule is often drawn ◯——— . The end shown by the circle will mix with water, the straight part 'prefers' to be in grease or oil.

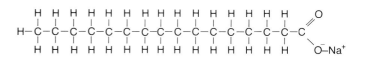

A sodium palmitate molecule

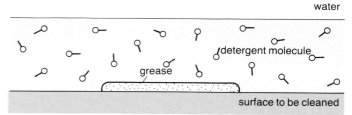

The grease and dirt is stuck to the surface to be cleaned The detergent has reduced the surface tension of the water so the surface is thoroughly wet.

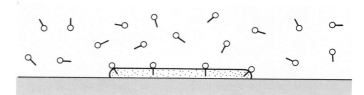

The hydrophobic 'tail' of the detergent molecule attaches itself to the grease. The hydrophilic 'head' stays in the water.

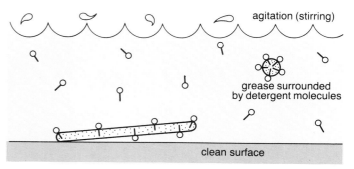

When the water is stirred or agitated the detergent molecules can get between the grease and the surface. The grease is lifted away forming a small globule surrounded by detergent molecules.

Making emulsions

Having got the dirt off our clothes, dishes or skin we want to make sure that it stays in the washing water and does not stick back on again. Detergent molecules help because they surround the small grease globules with their hydrophilic 'heads'. This helps to keep the grease suspended in the water.

These tiny pieces of grease are held in suspension. The detergent is acting as an emulsifying agent.

During washing it helps to stir or **agitate** the water. This lets the detergent molecules get between the dirt and the surface being cleaned. It also helps to break the grease globules into smaller pieces which can easily be held in suspension.

Microscopic globules of oil or grease suspended in water are called an **emulsion**. The photograph shows some of the emulsions we meet almost every day.

Examples of common, useful emulsions

Skin care and emulsions

Many of the skin care creams made by cosmetics companies are emulsions. When they are rubbed on to dry skin some of the water that the skin has lost is replaced. The oil which was in suspension forms a thin layer which makes the skin feel smooth. The greasy layer also stops any more evaporation from the skin.

Gardeners, factory workers, car mechanics and other workers often rub barrier cream into their skin before starting work in dirty conditions. This forms a soapy layer which protects the skin from dust, grease and dirt. After work the barrier cream can be washed away.

Barrier cream can be made in the laboratory.

Making barrier cream

Put on safety glasses.

*Take two beakers. In beaker **1** put 2 g of cetyl alcohol, 4 g of lanolin, 2 g of white petroleum (petroleum jelly), and 6 g of stearic acid.*

*In beaker **2** put 1 g of sodium hydroxide and 70 cm³ of water.* **Be careful. Sodium hydroxide is corrosive.**

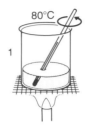

Heat both beakers gently. Use stirring thermometers to stir all the time and to check the temperatures.

*When the contents of each beaker are at 80°C, take them from the heat using suitable protection. Slowly pour the oils from beaker **1** into the hot sodium hydroxide solution. Be careful and stir all the time.*

When all the oils have been added and the mixture has been thoroughly stirred, add about 17 g of kaolin. Stir steadily until you have produced a creamy mixture. This is your barrier cream.

Lanolin is a natural grease from sheep's wool. Petroleum is a greasy substance obtained from oil. These form the oily part of the barrier cream.

Sodium hydroxide and stearic acid react to form sodium stearate. The molecules of sodium stearate stick into the globules of grease and keep them in suspension.

Problems with soap

The water which we use in our homes is not 'pure'. It has been in contact with the air and has probably soaked through the soil and rocks before reaching the local water reservoir. As a result, tap water contains dissolved chemicals. In small quantities these are not dangerous, indeed some help to keep us healthy. Others make the water taste better.

Some of the chemicals dissolved in water make it difficult for soap to foam. Different parts of the country contain different amounts of these chemicals. While on holiday you may have noticed that your soap does not give as much lather as it does at home. Alternatively you may have found that a small amount of soap gives so much lather that it is difficult to rinse away!

Water which gives a lot of foam with soap is called **soft water**. Water which does not give much lather is called **hard water**. When soap is used in hard water a **scum** forms. This scum is left around the bath or sink when the washing water drains away.

Investigating tap water

Place some tap water on to a clean watch glass. Put the watch glass over a beaker of hot water. Keep the water in the beaker hot until all the water in the watch glass has evaporated.

Look closely at the watch glass. See if there is a white, solid deposit on the glass.

Repeat the experiment with distilled water. Is there a deposit this time?

What does this tell you about distilled water?

Activities

Do you live in a hard water area?

Check whether or not you live in a hard water area by finding the answers to these questions:

1 When you use soap in the bath is a ring of scum left around the bath when the water is let out? Do not use bubble bath or bath salts in the water!

2 Look inside the kettle in your home. Does it have a coating of white solid?

3 Look at the taps in your home. Does the hot tap have a hard deposit on it? (Look where the water comes out.)

4 Look in the dishes under pot plants. Has the water left a hard deposit behind?

5 Look at the map on this page. Is your home town in a hard water area? (Use an atlas if it helps.)

If the answers to these questions are 'yes' then the water in your home is hard.

If the answer to these questions is 'no' then you either live in a soft water area or your house has a water softener! (See page 198.)

This map shows the amount of dissolved chemicals which cause hardness in water for different parts of the country. 200 parts per million makes the water noticeably hard.

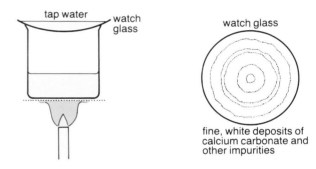

Investigating tap water

fine, white deposits of calcium carbonate and other impurities

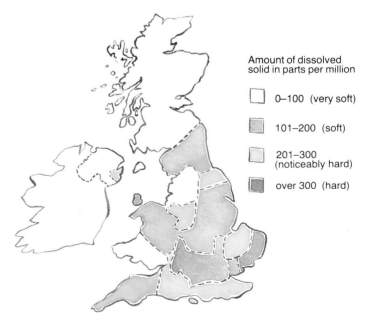

Amount of dissolved solid in parts per million

0–100 (very soft)

101–200 (soft)

201–300 (noticeably hard)

over 300 (hard)

Water hardness in the United Kingdom

Which chemicals cause hard water

Whether your tap water is hard or soft depends on the rocks it has been in contact with. In the British Isles some common rocks are limestone (mainly calcium carbonate with some magnesium carbonate), gypsum (calcium sulphate), rock salt (sodium chloride) and iron oxides. The ions which are likely to be in our tap water are Na^+ (sodium), Ca^{2+} (calcium), Fe^{2+} (iron), Mg^{2+} (magnesium), CO_3^{2-} (carbonate), SO_4^{2-} (sulphate) and Cl^- (chloride). We can find out which of these cause hard water by shaking solutions containing them with soap solution.

Calcium is the most common cause of hard water because much of our water runs through limestone rocks.

Limestone and hard water

We can think of limestone as calcium carbonate. Calcium carbonate is not very soluble in water but it does react with weakly acidic rain water (see page 83).

When the acidic rain water flows over limestone or chalk it reacts to form calcium hydrogencarbonate. This does dissolve in the water making it hard.

carbonic + calcium → calcium + carbon + water
acid   carbonate   hydrogencarbonate dioxide
          (doesn't dissolve)   (dissolves)

How does 'scale' form?

When hard water is heated, the calcium hydrogencarbonate dissolved in the water is decomposed. This means that it is broken down into calcium carbonate, carbon dioxide and water. The calcium carbonate is insoluble and so forms a hard, white deposit.

calcium   + heat → calcium + carbon + water
hydrogencarbonate   carbonate dioxide
   (in solution)   (solid)

This is the same reaction which produces the beautiful **stalactites** and **stalagmites** in limestone caves.

The warmth of the air in the cave decomposes some of the calcium hydrogencarbonate. The dripping water leaves a very tiny deposit of calcium carbonate. Over thousands of years this grows into a stalactite and, where the water splashes, a stalagmite.

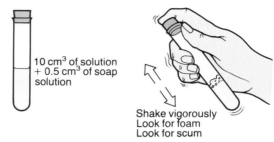

10 cm³ of solution
+ 0.5 cm³ of soap
solution

Shake vigorously
Look for foam
Look for scum

solutions	ions	observations
sodium chloride	Na^+ Cl^-	lots of bubbles
calcium chloride	Ca^{2+} Cl^-	scum and few bubbles
sodium nitrate	Na^+ NO_3^-	lots of bubbles
magnesium nitrate	Mg^{2+} NO_3^-	no lather; scum
sodium sulphate	Na^+ SO_4^{2-}	lots of lather
iron(II) sulphate	Fe^{2+} SO_4^{2-}	only a few bubbles; scum

Sample results

Advantages and disadvantages of hard water

The dissolved chemicals which make water hard can be of advantage:

1 Hard water usually tastes better than soft water. The label on the bottle of mineral water shows that it contains magnesium and calcium ions . . . but the water tastes good!

2 The dissolved calcium in hard water can help to produce strong teeth and healthy bones in children.

3 Where lead piping is used in old houses, hard water coats these with a layer of insoluble calcium carbonate. This stops any of the poisonous lead dissolving in the drinking water.

Against these advantages there are some serious disadvantages.

1 Hard water produces scum with soap. This looks unsightly on baths and basins. It also makes clothes washed in soap powder look dull.

2 Because soap does not lather well in hard water areas more soap is needed when washing. This can be expensive.

3 Some industries have to remove the hardness from water before using it. This is expensive.

4 Hard water forms 'scale' or 'fur' inside pipes and water heaters. The scale can block pipes which will then need cleaning or replacing.
'Fur' on heating elements makes them less efficient. This makes it more expensive to heat the water.

Making hard water soft

Softening by ion exchange

We have seen on page 197 that calcium ions in water cause hardness. The water can be made soft by 'exchanging' the calcium ions for sodium ions. We can see how this works in a dishwasher. If glasses were washed in a dishwasher using hard water they would dry with a white deposit of calcium carbonate. This would make them look dull and not clean. To avoid this, salt (sodium chloride) is poured into a special chamber in the washer. As the water for rinsing the glasses passes through the chemicals in the chamber, many of its calcium ions are replaced by sodium ions from the salt. This makes the water softer and so the glasses look much cleaner.

Where we want to soften all the water used in a house or factory we can fit an ion exchange column in the cold water supply. The column is a tube filled with a special chemical called a **resin**. The resin has lots of sodium ions. As the water flows through the column the calcium ions stay on the resin and the sodium ions go into the water. The water that comes out of the column is soft.

Using bath salts

Bath salts can be used to soften bath water. They work because they are made from sodium carbonate. When the bath salts are added to the bath water, calcium ions are removed as a thin deposit of calcium carbonate. This makes the water soft. As a result your soap gives more lather and you do not get left with a bath-tub ring.

Pure sodium carbonate is cheap to buy but not very attractive. Manufacturers of bath salts add dyes and perfumes. They then pack the 'bath crystals' in attractive boxes and glass jars. These things encourage customers to buy bath salts but do not make the water any softer!

Boiling and water softening

When hard water is boiled in a kettle, 'scale' or 'fur' is left inside (see page 197). The water poured from the kettle is much softer than water from the tap because it does not contain calcium hydrogencarbonate. Unfortunately the water still contains some calcium sulphate.

The hardness due to calcium sulphate cannot be removed by boiling so it is called **permanent hardness**. The hardness due to calcium hydrogencarbonate can be removed by boiling so it is called **temporary hardness**.

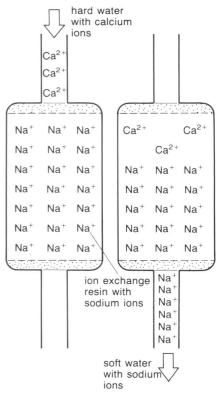

Softening water by ion exchange

Questions

1 Name two chemicals which can cause hardness in water.

2 What is temporary hardness?

3 Why is water softer after bath salts have been added?

4 'Bath crystals' bought for a present are much more expensive than the same weight of sodium carbonate. Why do you think this is?

5 A family lives in a house in a hard water area. Give three ways in which having hard water can cause extra expense.

6 A person fits a water softening device in the water supply.
    a) Suggest three advantages of having soft water.
    b) Suggest two disadvantages of having a water softener.

7 One chemical used in domestic water softeners has the trade name 'Calgon'. Why do you think that this name was chosen?

Synthetic detergents

Soap has been used for more than 2000 years but it does have its disadvantages. As scientists developed an understanding of how the soap molecule works, they tried to make substitutes for soaps made from fats and plant oils. These man-made cleaning substances are called **synthetic detergents**.

The first synthetic detergents were made in Germany during the First World War. This was so that the fats used for soap making could be used for other things.

After the war scientists throughout the world found that they could make detergents which were better than soaps! For example, in hard water soap forms a scum. The scum appears as a dirty ring around the bath or a whitish smear on glassware. Even after rinsing, hair washed with soap in a hard water area looks dull. Synthetic detergents do not form a scum.

Be careful! Concentrated sulphuric acid is very corrosive. Wear eye protection and do not spill the acid.

Making a synthetic (soapless) detergent

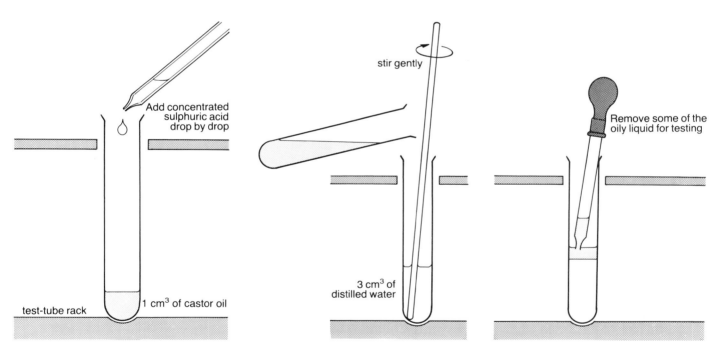

In the laboratory we can make a soapless detergent by reacting an oil with concentrated sulphuric acid.

Pour 1 cm³ of castor oil into a test tube. Stand the test tube in a rack.

Add about 2 cm³ of concentrated sulphuric acid to the castor oil.

Add the acid a few drops at a time and gently stir the mixture with a glass rod.

After all the acid has been added, keep stirring for two or three more minutes.

Slowly and carefully pour the contents of the test tube into a boiling tube containing 3 cm³ of distilled water. Stir gently for a few minutes and then leave the solution to stand.

Your soapless detergent will gradually separate as an oily liquid. Remove some of this for testing using a clean teat pipette.

Half fill a test tube with distilled water. Add some of the oily liquid you have prepared. Cork the test tube and then shake it. If you have been successful a foam of detergent bubbles will form.

Industrial preparation of soapless detergents

Manufacturers of the detergents we use in the home use chemicals from oil as their raw material. The ones chosen are hydrocarbons with a ring of carbon atoms joined to a long 'tail' of more carbon atoms.

The diagram shows the stages in preparing a synthetic detergent for sale.

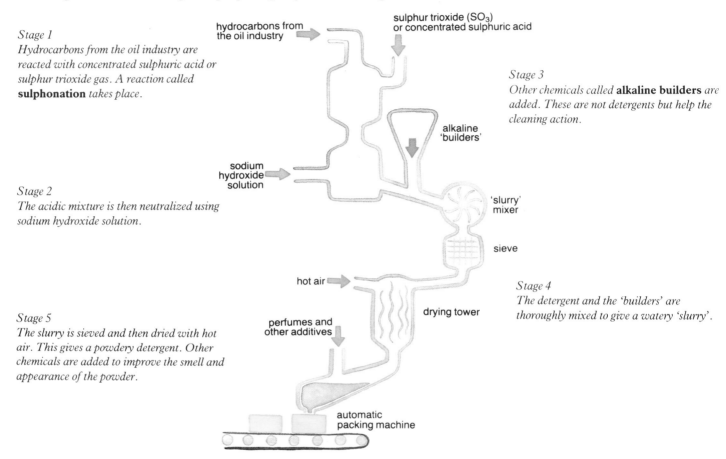

Stage 1
Hydrocarbons from the oil industry are reacted with concentrated sulphuric acid or sulphur trioxide gas. A reaction called **sulphonation** *takes place.*

Stage 2
The acidic mixture is then neutralized using sodium hydroxide solution.

Stage 5
The slurry is sieved and then dried with hot air. This gives a powdery detergent. Other chemicals are added to improve the smell and appearance of the powder.

Stage 3
Other chemicals called **alkaline builders** *are added. These are not detergents but help the cleaning action.*

Stage 4
The detergent and the 'builders' are thoroughly mixed to give a watery 'slurry'.

What do they add to make detergents 'wash whiter'?

Some manufacturers say that clothes washed in their detergent look 'whiter than white'. In a way this is true! When the washing powder is being made, chemicals called **optical brighteners** are added. These convert invisible ultra-violet light into light which we can see.

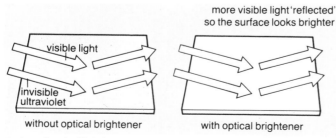

Action of optical brighteners

Getting rid of stains

Soaps and detergents are very good at lifting grease and dirt from cloth. There are, however, some stains which are difficult to remove. Things like egg, milk, and blood contain chemicals called proteins. They stick strongly to cloth and they do not dissolve in water. The detergent molecules cannot get underneath to lift the stain.

Some washing powders contain **enzymes** (see page 186) which can break down proteins. The enzymes make the stain soluble in water or at least allow the water and detergent molecules to get through to the cloth fibres.

Enzymes are chemicals which control reactions in living organisms. Because of this, washing powders with enzymes are often said to have 'biological action'.

Detergents in the environment

Synthetic detergents are sometimes called soapless detergents. They have some advantages over real soaps, particularly in hard water areas. Because of this they are used in very large quantities in homes, laundries and in industry. Waste water from these places goes to sewage treatment plants where bacteria are used to break down the waste material into harmless chemicals. After this the water is perfectly safe to run off into rivers or even reservoirs.

In the 1950s it was found that waste water containing synthetic detergents was making lots of foam at the sewage plant. The bacteria could not break this down and so great masses of foam were released into rivers.

The foam, sometimes two metres high, did not look nice but more seriously, it killed plants and animals in the river. As these rotted away, long stretches of river became heavily polluted. Eventually all the fish in these rivers died.

Soap foam is **biodegradable**. This means that bacteria can break it down into harmless gases. The foam from early synthetic detergents was **non-biodegradable**.

Scientists began to search for a soapless detergent molecule which could be attacked by bacteria.

It had a 'head' and a 'tail' like a soap molecule but the tail was not one straight line of carbon atoms. It had some branches and this shape stopped it from being broken down by bacteria.

Scientists then 'designed' a soapless detergent molecule with a shape more like that of soap. The diagram below shows the structure of a modern detergent. It is made from a petroleum product.

The foam from the detergent **is** biodegradable. This means that the sewage discharged into our rivers does not contain harmful detergents. As a result, many of our rivers, including the Thames, have completely recovered. Rivers running through our cities may now be cleaner than they have been for hundreds of years.

Questions

1   What does 'biodegradable' mean?

2   What do bacteria do to soap foam in a sewage plant?

3   What problems did early non-biodegradable detergents cause?

4   What was wrong with the shape of early detergent molecules?

5   Some plastics are non-biodegradable. What problems do these plastics cause in the environment?

6   Scientists can make plastics which can be 'degraded' by sunlight. How would these help to keep our countryside clean?

7   Nature conservation groups campaign to make sure that factories do not discharge dangerous waste into rivers. Why is this important for people who want to visit the countryside?

8   Why do you think that rivers through large towns were polluted hundreds of years ago?

Food from fats

Before the First World War, all the detergents used were soaps made from animal fat or plant oil. The demand for detergents grew very quickly and at the same time it became important to conserve food supplies. These things led scientists to work on the development of soapless detergents. This meant that some of the fats and plant oils which had been used for soap could be turned into valuable food such as **margarine**.

The oil for making margarine comes from the seeds of plants such as sunflowers. The oil contains complicated molecules called organic acids joined to a glycerol molecule. Some of the organic acids are **saturated**. All their carbon atoms are joined together by single bonds (C—C) and so they cannot take any more hydrogen atoms. Other acids are **unsaturated**. Here some of the carbon atoms have formed double bonds (C=C). These double bonds can be broken and joined to hydrogen atoms. The new, saturated molecule will contain more hydrogen than the unsaturated molecule.

$$H-\overset{\displaystyle H}{\underset{\displaystyle H}{C}}-\overset{\displaystyle H}{\underset{\displaystyle H}{C}}-\overset{\displaystyle H}{\underset{\displaystyle H}{C}}-\overset{\displaystyle H}{\underset{\displaystyle H}{C}}-\overset{\displaystyle H}{C}=\overset{\displaystyle H}{C}-\overset{\displaystyle H}{\underset{\displaystyle H}{C}}-\overset{\displaystyle H}{\underset{\displaystyle H}{C}}-\overset{\displaystyle H}{\underset{\displaystyle H}{C}}-COOH$$

*An **unsaturated** organic acid*

The plant oil can be made into a solid fat by making the unsaturated acids take up more hydrogen until they are saturated. This is called **hydrogenation**.

$$H-\overset{\displaystyle H}{\underset{\displaystyle H}{C}}-\overset{\displaystyle H}{\underset{\displaystyle H}{C}}-\overset{\displaystyle H}{\underset{\displaystyle H}{C}}-\overset{\displaystyle H}{\underset{\displaystyle H}{C}}-\overset{\displaystyle H}{\underset{\displaystyle H}{C}}-\overset{\displaystyle H}{\underset{\displaystyle H}{C}}-\overset{\displaystyle H}{\underset{\displaystyle H}{C}}-\overset{\displaystyle H}{\underset{\displaystyle H}{C}}-\overset{\displaystyle H}{\underset{\displaystyle H}{C}}-COOH$$

*A **saturated** organic acid*

Making margarine

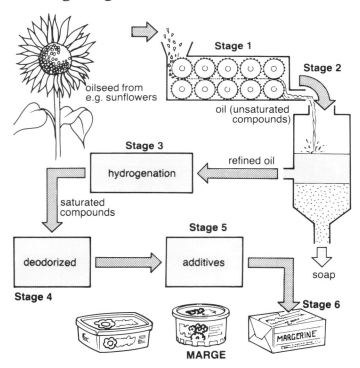

Stage 1
The sunflower seeds are crushed and squeezed to extract oil.

Stage 2
The oil is refined by heating it with sodium hydroxide. The impurities in the oil react to form a sort of soap. The purified oil is separated and then washed.

Stage 3
Hydrogenation takes place when the oil is heated with hydrogen under pressure over a nickel catalyst.

The catalyst speeds up the reaction which would otherwise be very slow.

Stage 4
The fat from *Stage 3* has an unpleasant smell and would not attract many buyers! It is heated and then has steam blown through it. This takes away the odour.

Stage 5
Most people use margarine as a substitute for butter. To make margarine more like butter, colouring, flavouring and salt may be added. Vitamins are also added to make it healthier to eat.

Stage 6
The margarine is finally packed for sale. Hard margarines are wrapped in paper. Softer margarines are packed in plastic tubs.

1 The structure of a soap molecule is shown below. The soap is called sodium palmitate.

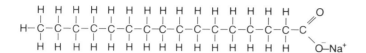

a) Is this soap made from animal fat or plant oil? (The name gives you a clue!)

b) Concentrated sodium hydroxide solution is added to make the soap. Describe the safety precautions you should take when using concentrated sodium hydroxide solution. Why are safety precautions necessary?

c) How would you show someone who was not a scientist that soaps and detergents reduce the surface tension of water?

2 **a)** State two advantages of synthetic (soapless) detergents over soap.

b) A detergent molecule can be drawn

i) How do the two parts of the molecule behave in water?

ii) A garage mechanic's hands are covered in engine oil. Use diagrams to show how washing with detergent will remove the oil from the mechanic's skin.

3 **a)** How can the owner of a house find out if water from the taps is hard or soft?

b) Water pipes in old houses were made from lead. Lead is a poisonous metal which can build up in the body. Even very small quantities of lead taken regularly can cause sickness and death.
Doctors have found that people living in old houses in Glasgow are more likely to show signs of lead poisoning than people from London.
Suggest why this is. (The map on page 196 may help.)

4 The label below is from a bottle of mineral water. It shows that the water contains the following ions:

CO-OP GLENDALE SPRING
CARBONATED NATURAL MINERAL WATER

CONSUMER CARE
PRODUCT INFORMATION

Co-op GLENDALE SPRING Carbonated Natural Mineral Water, produced with water drawn from the Glendale Spring, Perthshire, Scotland. This calorie free water is delicious served chilled, or mixed with Fruit Juice for added refreshment.

Composition in accordance with the results of the officially recognised analysis of 9 June 1987.
Contains (mg/l)

Bicarbonates	156	Calcium	28.8
Sulphates	10.3	Magnesium	9.6
Chlorides	29.0	Fluorides	0.10
Nitrates	0.24	Sodium	32.0

pH 7.4

CO-OPERATIVE WHOLESALE SOCIETY LTD., NEW

a) Which of the ions listed cause hardness?

b) If the mineral water was boiled what type of hardness would be removed?

c) This mineral water comes from a spring. Nearby there are some caves with stalagmites and stalactites.

i) What is the difference between a stalagmite and a stalactite?

ii) How does a stalactite form?

5 The advert below is for a new washing powder.

New, Improved **Sudso** washing powder
Our new formula gives you a wash that makes even your brightest whites look whiter! Even those difficult stains like egg and blood come out easily. No need for soaking; new bio-action **Sudso** does it all!
And after the wash, whether your water is hard or soft **Sudso** leaves your washing machine scum-free and sparkling clean.

*By hand or machine, just **Sudso** it clean*

a) Is the washing powder a soap or a synthetic detergent? How do you know?

b) Explain how added chemicals can make white things look whiter?

c) What do you think has been added to the washing powder to help remove the stains listed?

6 An advert for 'Kitchen Wizard' washing-up liquid says:

'Kitchen Wizard cleans more plates than any other washing-up liquid for the same cost.'

Describe how you could check this claim by a scientific test. Give full details of how you would make sure that the test was properly controlled.

What is 'weather'?
What causes the weather?
Can we predict tomorrow's weather?

SNOWSTORM BRINGS DOWN PYLONS – TWO KILLED

BUXTON CUT OFF IN BLIZZARD

Ethiopia – drought conditions continue

OCTOBER GALES WORST FOR 200 YEARS

Torrential rain sweeps West Country – many casualties

Southern England bakes in hottest spell since 1976 – hosepipes banned

The weather is very important to us. It affects how we spend our time, what we choose to wear and even how healthy we feel. For some people the effects of bad weather may be very serious. Farmers may lose their crops if it is too wet or too cold, fishermen may be shipwrecked in stormy weather and, in certain countries, hundreds of people may die because of floods, droughts or hurricanes.

The study of the weather is called **meteorology**. Meteorologists study and record weather patterns over very long periods of time. This helps them to understand what causes the weather. Using readings from weather stations all over the world, measurements taken by weather balloons and photographs taken from satellites, weathermen can tell, to a certain extent, what the weather will be like tomorrow.

Accurate weather forecasting allows people to prepare for bad weather. For example, when snow is likely local councils can grit and salt roads. In countries where hurricanes occur, early warnings allow people to protect their property and to get to safety. Gale warnings are essential for those at sea so that they can prepare for storms or head for the nearest harbour. A forecast for shipping is broadcast regularly on BBC radio.

Activities

1 Make a list of words connected with **bad** weather. Three are given to start you off. Floods, drought, hurricane, . . .

2 Make a list of the good things that could happen if scientists could control the weather. For example, 'It could be arranged so that the weather was always good at weekends and on public holidays'.

3 Cut out and keep all the newspaper reports you can find concerning the weather. Do this for about three weeks and then make a poster showing how the weather has affected the world. (*Hint*: do not just look on the front pages.)

What is the 'weather'?

Our planet is surrounded by a layer of air called the **atmosphere**. The atmosphere is about 160 km thick but our weather occurs in the 16 km closest to the Earth's surface. In this layer of air conditions vary from place to place. For example, above the Sahara desert it may be hot and dry but above the North Sea it is usually cold and wet.

The weather describes what it is like at a particular place in the atmosphere at a particular time.

This table was printed in a newspaper. It shows the weather conditions at various places around the world. The letters (S, C, R, Fg) tell us whether a place was Sunny, Cloudy, Rainy or Foggy. The number in the next column gives the temperature in degrees Celsius.

Questions

1 What is the hottest place listed? Find that place on a map. What is the coldest place listed? Find that place on a map.

2 Find all the English towns listed. (There are four.) Find the temperature difference between the hottest and the coldest.

The gases in the atmosphere are always moving because energy from the Sun heats some parts more than others. As the air moves, the weather conditions change.

The movement of gases in the atmosphere is very complicated but, by recording the weather for many days, we can find **patterns**. These 'patterns' tell us what usually happens. For example, winds from the north usually, but not always, bring cold weather to the United Kingdom. Similarly, July is usually, but not always, the hottest month in England.

These patterns help us to predict what the weather will be like in the future. Newspapers, radio, and television give weather forecasts each day.

Activities

1 Listen to the weather forecast each evening for seven days. During the next day check how accurate it was. Record your results in a table with the headings: Date, Weather forecast, Actual weather. Do you think that the forecasts are 'nearly always right', 'sometimes right', or 'nearly always wrong'?

2 Use weather forecasts from television, radio, or a newspaper to find out what the weather is like in different parts of the United Kingdom. Each day make a note of the weather in Scotland, Wales, south-west England (Devon and Cornwall), and East Anglia. After you have collected the information for about ten days check whether these statements are true.

'It is usually colder in the north than in the south.'

'It is usually wetter in the west than in the east.'

'The south west has more sun than other parts.'

OUT AND ABOUT
MIDDAY: S=sun; C=cloud; R=rain; Fg=fog.

Place		C	Place		C
Ajaccio	s	28	Majorca	s	30
Akrotiri	s	28	Malaga	s	31
Alexandria	s	29	Malta	s	30
Algiers	s	32	Melbourne	c	11
Amsterdam	c	21	Manchester	c	14
Athens	s	31	Mexico City	–	–
Bahrain	s	36	Miami	s	30
Barbados	s	30	Milan	s	29
Barcelona	c	29	Montreal	–	–
Belgrade	c	30	Moscow	c	20
Berlin	s	23	Munich	c	21
Bermuda	–	–	Nairobi	s	21
Biarritz	c	23	Naples	c	33
Birmingham	c	14	Newcastle-upon-Tyne	c	15
Bordeaux	c	24	New Delhi	s	35
Brussels	c	23	New York	s	22
Budapest	c	29	Nice	s	27
Buenos Aires	–	–	Oslo	c	17
Cairo	s	33	Paris	c	23
Cape Town	–	–	Peking	s	33
Casablanca	c	25	Perth	s	17
Chicago	c	23	Prague	s	22
Christchurch	c	12	Reykjvik	c	11
Cologne	s	23	Rhodes	s	30
Copenhagen	c	18	Rio de Janeiro	s	21
Corfu	s	32	Riyadh	s	43
Dublin	c	20	Rome	s	28
Dubrovnik	s	28	Salzburg	c	24
Faro	s	27	San Francisco	s	21
Florence	s	30	Santiago	s	15
Frankfurt	c	21	San Paulo	c	20
Funchal	s	26	Seoul	s	30
Geneva	r	20	Singapore	–	–
Gibraltar	s	28	Stockholm	c	21
Helsinki	c	22	Strasbourg	c	22
Hong Kong	s	32	Sydney	c	13
Innsbruck	c	19	Tangier	s	27
Istanbul	s	28	Tel Aviv	s	30
Jeddah	s	39	Tenerife	s	27
Johannesburg	s	18	Tokyo	s	31
Karachi	–	–	Toronto	c	18
Las Palmas	s	27	Tunis	s	33
Le Touquet	c	20	Valencia	s	30
Lisbon	s	27	Vancouver	–	–
Locarno	c	27	Venice	c	26
London	s	21	Vienna	c	21
Los Angeles	c	21	Warsaw	c	24
Luxembourg	c	20	Washington	s	24
Luxor	s	42	Wellington	s	13
Madrid	s	31	Zurich	c	21

Weather forecasts are given in newspapers, on radio, and on television. How accurate are they?

Energy from the sun

The energy which causes the changing weather on our planet comes from the Sun. The Sun is a star which generates huge amounts of energy in each second. This is **radiated** in all directions.

The energy is in the form of **electromagnetic** waves. These travel across the vacuum of outer space at a speed of about **300 000 000 metres per second**!

Although all the waves from the Sun travel at the same speed, they have a wide range of wavelengths. The diagram below shows the types of radiation from the Sun. The longest are radiowaves and the shortest are called gamma rays.

How many of these types of waves do you recognize? Can you say what they are used for?

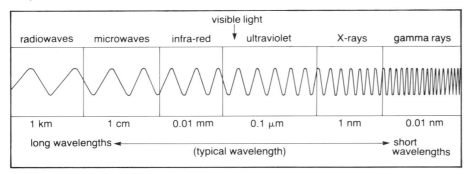

The electromagnetic spectrum (Not drawn to scale)

1 km	=	1000 m
1 cm	=	0.01 m
1 mm	=	0.001 m
1 μm	=	0.000001 m
1 nm	=	0.000000001 m

Most of the long waves from the Sun are reflected or absorbed by the upper and middle layers of the atmosphere. Some of the short waves do get through to the Earth's surface. The energy carried by these waves is absorbed by the Earth causing it to warm up. This, in turn, heats up the air nearby.

The air near the Earth's surface is warmer than that higher up.

What types of surfaces are best at absorbing radiation?

The diagram shows a heater producing infrared radiation which is good for heating substances. The metal plates near the heater have coins stuck to them with wax. One plate is polished and the other is painted dull black.

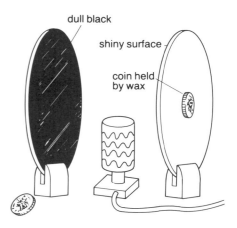

Investigating absorption

The coin falls from the black plate first showing that it has absorbed enough energy to melt the wax.

Dull, dark surfaces are good absorbers of radiation. Shiny, light surfaces reflect radiation and so are poor absorbers.

When energy from the Sun reaches the Earth, it can be reflected or absorbed. Brown soil is a good absorber of radiation and so heats up quickly. Snow and water are good reflectors.

Radiation and the atmosphere

Only a small fraction of the Sun's radiated energy reaches the Earth's atmosphere. Of this, only about 45 per cent gets to the layer of air near the Earth's surface where the weather forms. The rest is either reflected by the upper layers of the atmosphere (40 per cent) or absorbed by dust, water vapour and gases.

Notice from the diagram below that much of the ultraviolet radiation is absorbed by a layer of **ozone gas**.

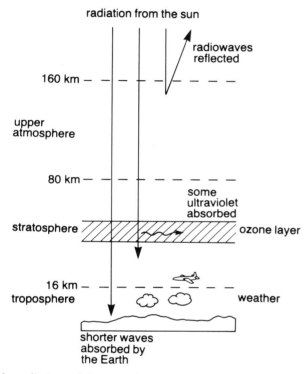

Solar radiation and the atmosphere

Questions

1 The Sun is about 150 000 000 km from the Earth. Electromagnetic waves travel at a speed of 300 000 km/s.
Roughly how long does it take energy to reach us from the Sun?

2 Draw a sketch to show the Sun radiating energy in all directions. Use your sketch to explain why only a small fraction of the Sun's energy reaches the Earth's atmosphere.

3 The climber in the photograph gets warm but the white snow does not melt. Why?

4 What colour suit would keep the climber warmest on a sunny day?

5 The climber may get a sun tan. Find out what type off radiation causes this. What does this tell you about the ozone layer?

Radiation which does reach the Earth meets lots of different surfaces. Some are very reflective, others absorb much of the radiation which falls on them. The table shows the percentage of the radiation reflected and absorbed for some common materials.

substance	percentage reflected	percentage absorbed
forests	5–10	90–95
dry earth	10–25	75–90
sand	20–30	70–80
grass	20–30	70–80
clouds	50–60	40–50
snow	75–85	15–25

Measuring temperature

One of the most important pieces of information given in a weather report is the temperature.

Scientists, including meteorologists use the **Celsius scale**. On the Celsius scale, the temperature of melting ice is 0°C. The temperature of boiling water is 100°C.

The temperature on our planet varies a great deal from place to place. According to modern weather records, the highest temperature ever was 58°C in Libya. The lowest temperature recorded was −89.2°C in Antarctica.

Activities

1 Using a thermometer marked in degrees Celsius (°C), measure and record the temperatures of the following: the air in a room such as a school laboratory, the air outside, water taken from a tap, water taken from a tap and allowed to stand in a room for two hours, your skin temperature.

2 We often think that our skin is a very good thermometer. The following experiment shows that it is not that simple!

Take three bowls big enough to put your hands in. Place them side by side.

Fill the first bowl with iced water at about 5°C. Fill the middle bowl with water from the tap at about 15°C. Fill the last bowl with warm water at about 30°C.

Now put one hand in the cold water and the other in the warm water. Keep them there for about two minutes. After this time, place both hands in the middle bowl. What do you feel?

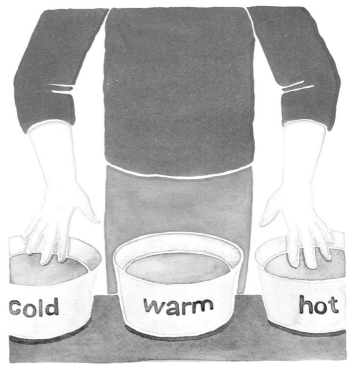

Maximum and minimum thermometers

The photograph shows a special type of thermometer called a 'maximum and minimum' thermometer. It works just like a normal thermometer in that the liquid (mercury) level shows the temperature but, as it moves up and down, the mercury pushes two little markers. The one on the right is pushed up as the temperature rises but is left behind as the temperature falls again. The marker shows the highest temperature of the day (maximum). The other marker is moved as the temperature falls and is left to show the lowest temperature of the day (minimum).

Questions

1 Gardeners often use a maximum and minimum thermometer in a greenhouse. Why is this necessary in **a)** winter **b)** summer?

2 The markers in maximum and minimum thermometers have to be reset to their starting positions every day. Why is this?

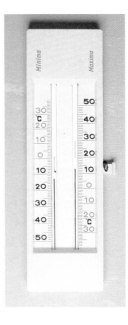

A maximum and minimum thermometer (There are different types.)

Why is it cold at the North Pole?

The north and south poles of our planet are covered in snow and ice and the temperatures there are always low. In contrast, the lands near the equator are hot. Some are deserts and others have hot, humid jungles. Why is it much hotter at the equator?

We can get part of the answer by drawing a diagram of our planet showing how energy from the Sun reaches the Earth.

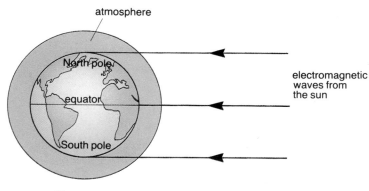

(Not to scale)

Look at the diagram. See how much air the rays travelling to the poles have to pass through. This air absorbs energy and so not much is left to heat up the Earth. At the equator, the rays pass through a much thinner layer of air and so have more energy left to heat up the land.

This is only part of the answer. Imagine thick 'beams of energy' from the Sun reaching the Earth as shown on the diagram below.

When beam 1 reaches the Earth, its energy heats up a small area of land. However, because our planet is shaped like a ball, beam 2 is spread over a larger area and so does not have such a big heating effect.

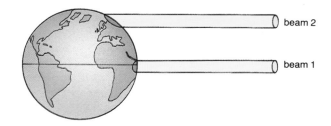

Activities

Fix a torch so that it gives a beam of light parallel to a table or bench top.

Put a large book on the table in front of the torch and then lift up the cover so that you can see a circular patch of light on it.

Now slowly lower the cover of the book so that the angle shown on the diagram gets smaller. What happens to the area covered by the patch of light?

How does this result help to explain why there is ice at the north pole but not at the equator?

This quick observation can be turned into a careful experiment. How could you measure the area covered by the light at different angles?

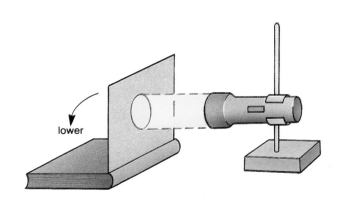

Water in the atmosphere

The photograph opposite was taken from a satellite and shows one side of our planet. (Europe and the other continents are on the other side and so are hidden.) The dark areas are the oceans and the light patches are clouds. It shows that we live on a very watery planet. In fact only about one third of the Earth's surface is dry land.

By taking photographs like these at regular intervals, meteorologists can tell where the clouds are, where they are moving to and at what speed. This allows them to predict where and when rain will fall. Look out for these 'satellite pictures' on television weather forecasts.

Evaporation

Energy from the Sun evaporates some of the water from oceans, seas, rivers and lakes. The energy makes the particles in the water move faster. Some of the particles get enough energy to break free of the liquid's surface. These form an invisible gas called water vapour. Evaporation is when a liquid turns to a gas without being heated to its boiling point.

Water vapour is one of the gases of the atmosphere. The table opposite shows the composition of the atmosphere. The percentages for nitrogen, oxygen, and carbon dioxide vary slightly but the percentage of water vapour varies a great deal. The amount of water vapour in the air is an important part of the weather.

Planet Earth

gas	percentage volume
nitrogen	about 80%
oxygen	about 20%
carbon dioxide	about 0.03%
noble gases	about 1%
water vapour	between 0% and 5%

Activities

Take four saucers or flat dishes and place the same amount of water in each. Place one near a radiator and another in a cool place. (You could use a refrigerator.) Now use plastic wrapping film to cover partially the tops of the remaining containers. (Leave about one quarter uncovered.) Place one of these near to a radiator and the other in the cool place.

Each day, look at the containers. See how much water is left. What does this experiment tell you about evaporation?

Questions

1 During the summer a bird bath needs topping up every day. In the autumn it needs topping up once a week. One reason is probably that the birds drink more in the summer but can you think of another reason?

2 Central heating makes the air in a house very dry. This is thought to be bad for wooden furniture. How could an open container of water hanging on a radiator help?

3 The humidity (amount of water vapour in the air) is usually very low in the middle of the desert. It is also very low in the Arctic. Suggest why these two places have low humidity.

Clouds by convection

We have seen how clouds form when warm, moist air is forced to rise over hills or mountains. However, clouds can form in the middle of large, flat land masses hundreds of miles from the nearest sea. These clouds form by **convection**.

On a warm day, water in lakes and rivers will evaporate. Water in the soil also evaporates. In addition, the transpiration of plants (see Chapter 6) releases water vapour into the atmosphere. All these things produce a volume of warm, moist air near to the Earth's surface.

This warm air is less dense than the cool air above and so the denser air falls, pushing the warm air upwards. This falling of cool air and rising of warm air is called a **convection current**.

As the 'bubble' of warm air rises, the atmospheric pressure around it gets less and so the 'bubble' expands. This expansion cools the air at the rate of about 1 °C for every 100 m rise.

If the rising air cools enough, there will be a certain height at which the temperature allows the water vapour to condense. A cloud will form with its base at this height.

The photograph shows a **cumulonimbus** cloud. This is the type of cloud which usually brings thunder storms. It is a towering cloud which builds up quickly as strong convection currents carry more and more water vapour into it. The flat top of the cumulonimbus cloud makes it look something like an metal worker's anvil.

In tropical regions such clouds form nearly every day as energy from the Sun makes water in the soil evaporate. This is then carried upwards on a current of warm air. Eventually the water falls back to earth as a tropical rain storm.

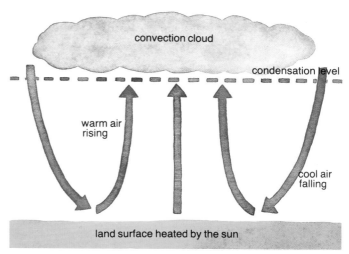

Convection cloud formation

Convection currents and flight

The upward current of warm air which causes convection clouds is sometimes called a **thermal**. The air in a thermal can rise at very great speeds. This can be of great use to birds (and humans) who fly by gliding because they can use the thermal to gain height without using any of their own energy.

Pilots of light aircraft try to avoid flying through convection clouds because the air inside the cloud is turbulent. This means that it is swirling around at great speeds. This can make the aircraft lose control.

Questions

1 Why do convection clouds tend to have flat bases?

2 Vultures are heavy birds with large wings for gliding. Suggest why vultures do not fly in the early morning but fly for longer periods of time during the middle part of the day.

3 Glider pilots like to find thermals so that they can fly higher. (Do not forget that they do not have an engine!) Suggest why glider pilots circle over motorways on hot days. (*Hint:* think of the road surface.)

4 Air gets about 1 °C cooler for every 100 m it rises. If the air just above the ground has a temperature of 21 °C and the 'dew point' is 2 °C, estimate the height at which clouds will start to form.
(The dew point is the temperature at which the air becomes saturated with water vapour.)

Rain, snow, and hail

When water vapour in a cloud condenses, the drops of liquid may be too heavy to stay suspended in the cloud. They then fall as rain.

In a cloud which is high up, the temperature is low and so the water turns to ice. Tiny ice crystals join together and may fall from the cloud as snowflakes. The photographs show some snow flakes. There are many different patterns but that they all have a roughly hexagonal (six-sided) shape.

The bottom of a cloud is warmer than the top. This causes convection currents inside the cloud. (See page 213). If the upward air currents are strong enough, drops of water at the base of the cloud are lifted up to the top where they turn to ice. The ice may then fall down only to be pushed up to the top of the cloud again and again. Each time, another layer of ice is frozen on until a large **hailstone** is formed. This then drops from the cloud.

Hailstones can cause damage to houses and to farmers' crops. Some hailstones are up to 10 cm in diameter!

Mist, fog, rain, snow and hail are all due to water vapour condensing from the atmosphere. Meteorologists call all these things **precipitation**.

Measuring rainfall

Rainfall is measured by collecting rain in a rain-gauge. A typical rain-gauge is shown below.

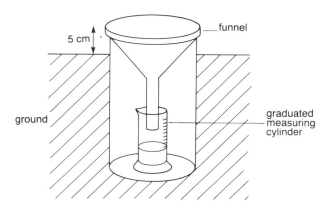

Rain gauge

Hail and snow have to be melted before being accurately measured but 10 mm of snow is equal to about 1 mm of rain.

Snowflakes

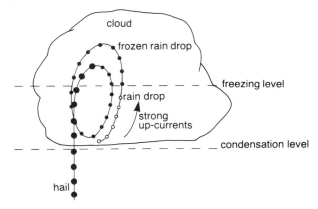

How hailstones form

Activities

Construct a rain-gauge. The funnel can be made by cutting the top from a plastic lemonade bottle. Cheap, plastic measuring cylinders can be bought at chemists' shops and the whole thing can be mounted in a tin can.

If you want to measure the rainfall in millimetres, you will need to mark your measuring cylinder with a scale. (**The scale marked on it is for measuring volume and so will not do.**)

The marks on your scale will need to be equally spaced. The distance between the marks should be calculated by:

$$\text{scale spacing} = \frac{D \times D}{d \times d} \text{ mm}$$

Where D = diameter of funnel
and d = diameter of measuring cylinder.

Can you see why a *wide* funnel and a *narrow* measuring cylinder are used in this gauge?

Rainfall patterns

The weather tells us about conditions at a certain place at a certain time. Measurements of weather conditions are taken regularly and recorded for many years. By looking for typical weather patterns the **climate** of a place can be found. **The climate tells us about typical or average weather conditions.**

Weather measurements are often recorded on graphs or bar charts. The bar charts below show the amount of 'precipitation' (rain, snow and hail) for each month of the year for three different places. The places are shown on the map.

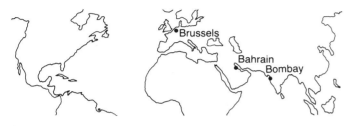

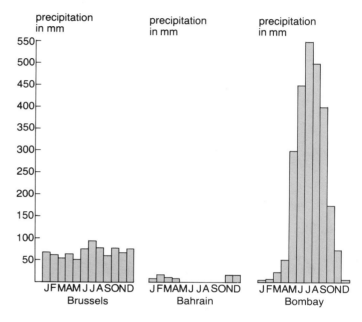

Questions

1 Which of the places is in the desert?

2 May to September is the monsoon season in India. What do you think that the word 'monsoon' means? (*Hint:* look at the chart for Bombay.)

3 Which place has the same sort of climate as London?

4 What roughly is the total annual rainfall for Brussels?

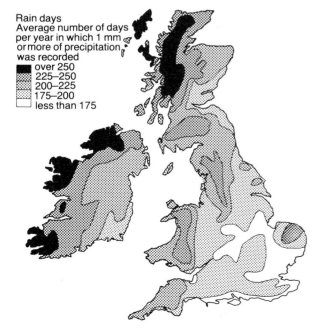

Rainfall in the U.K.

Rainfall in the United Kingdom

The map opposite shows the number of rainy days in different parts of the United Kingdom and Ireland. It is clear that the rainfall varies greatly from one part to another. We can explain this using our scientific knowledge of how clouds form. For example, we know that rain often falls when clouds rise over mountains. This explains why rainfall is often higher on one side of a mountain than the other.

Questions

5 Roughly how many rainy days does your home town have in a year?

6 Why do you think that the west coast of Ireland has so much rain? (*Hint:* what is between Ireland and America?)

7 Why do you think that the centre of Wales is so wet?

Pressure of the atmosphere

The atmosphere is a layer of air which is about 160 kilometres (100 miles) thick! Even though air is a gas, the atmosphere has a great weight which pushes down on us.

The square drawn opposite has an area of 1 square centimetre (1 cm²). Above it there is about 160 km of air pressing down with a weight of about 10 N. This is the weight of a 1 kg bag of sugar.

The force due to the air pushing down on this area is about 10 N.

We can say that the atmosphere exerts a pressure of about 10 newtons per square centimetre (10 N/cm²).

When studying the weather, we measure the pressure in units called millibars (symbol mb). Normal atmospheric pressure is 1000 mb.

Measuring air pressure

A rotating drum barometer records the air pressure on a drum which slowly revolves. The main part of the barometer is a flexible metal can. If the air pressure increases the can is squashed slightly and this moves the pen on the chart upwards. If the air pressure gets less the can expands slightly and the pen moves down.

The chart shows how the air pressure varied during one day.

Aneroid barometer (rotating drum)

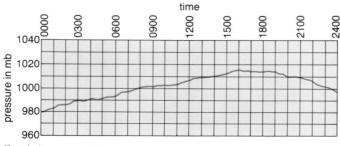

Graph from a recording barometer

Pressure and weather

In general, high pressure brings settled weather. In the summer we want to hear the weatherman say, 'There is a high pressure region over the United Kingdom. This anticyclone is likely to remain for some time'.

Low pressure usually brings strong winds and unsettled weather. In the winter we do not want to hear the weatherman say, 'There is a deep depression centred over the United Kingdom. This is likely to bring strong, cold winds from the north'.

For fishermen and others at sea it is very important to know whether the pressure is falling; a sudden drop in pressure may mean that a storm is on the way. Each day, weather forecasts are broadcast on the radio for shipping. The information is always given in the order listed below:

Place (You may need a special map if you want to find the places named.)

Wind direction (The direction the wind is coming from.)

Pressure (Given in millibars followed by whether it is rising, falling, or steady.)

Wind strength (Given on a scale where 0 means 'calm', 1–5 means 'light winds', 6–7 means 'strong wind', 8–9 means 'gale', 10–11 means 'storm' and 12 means 'hurricane'.)

Precipitation (Rain, hail, fog, snow.)

Visibility ('Good' means that you can see for a long way.)

Air pressure and wind direction

When you blow up a balloon and hold the neck tightly closed, the air inside the balloon is at a higher pressure than the air outside. If you now release the neck, the air rushes out of the balloon like a gust of wind. **In our atmosphere winds occur when air moves from a high pressure region to a low pressure region**.

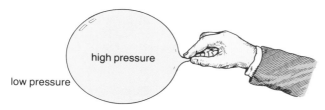

The pressure is not the same at all places in the atmosphere because some parts heat up faster than others. The air that heats up more expands and becomes less dense. This gives a low pressure region. Colder air is more dense and so gives a high pressure region.

This can be understood by thinking about land near to the sea.

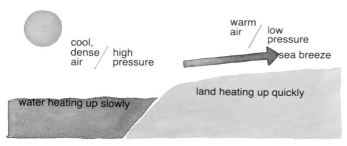

Sea breeze during the day

During the day, the land absorbs energy from the Sun faster than the sea does. This heats the air above which expands giving a low pressure region. The air above the sea is at a higher pressure and so a light wind or breeze blows in towards the land.

Questions

1 During the night, the land cools down faster than the sea.

   a) Do you think that the air over the sea will cool down faster than the air over the land?

   b) Will the air over the sea become more or less dense than the air over the land?

   c) Where will the pressure be higher, over the land or over the sea?

   d) Which way does the breeze blow during the night?

Pressure variation with height

The winds and breezes which are part of our weather described above occur in the lowest part of our atmosphere. They were caused by pressure differences due to heating effects. There is however, another pressure variation to be considered.

As we move higher in the atmosphere, for example when we climb a mountain, there is less air above us to push down. This means that the pressure gets less. This would be true if our atmosphere was uniform but is even more noticeable because the air gets 'thinner' as we go higher. More accurately, the air is less dense; the air molecules are further apart and so there are less in any given volume. For example, mountaineers climbing the world's highest peaks need to carry a supply of oxygen. Without it their brains and muscles would not work properly. This is because a lung full of air would not contain enough oxygen molecules for the body's needs.

The diagram shows how the pressure varies with height above the Earth's surface.

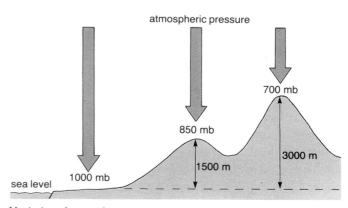

Variation of atmospheric pressure with height

2 Why do passenger aircraft need to be 'pressurized' whilst flying at great heights?
How do you think an aircraft is 'pressurized'?

3 The altimeter of an aircraft is the instrument used to measure the aircraft's height above the ground. It is actually a barometer.

   a) Why can a barometer be used as a 'height measurer'?

   b) Why does a pilot need to know the height above sea level of the airport at which he is to land?

Plotting pressure variations

We have seen that an atmospheric pressure measurement is useful in weather forecasting but we can get more information by recording the pressure at a large number of places and then looking for 'patterns'.

Pressure readings are taken from a large number of weather stations and then plotted on a map. Lines are then drawn to show where the atmospheric pressure has the same value.

These lines are called **isobars**. You can see isobars on the map opposite. The numbers on the lines give the pressure in millibars. (Remember, around 1000 mb can be taken as 'normal'.)

Isobars and wind direction

On the map you can see a high pressure region (also called an **anticyclone**) and a low pressure region (sometimes called a **depression**).

From our earlier work we would expect the wind to blow directly from the 'high' to the 'low' but it does not. This is because our planet spins. This affects the wind direction so that the winds blow roughly anticlockwise around depressions in the northern hemisphere and clockwise in the southern hemisphere.

The result, seen on the weather map, is that the winds tend to blow more or less parallel to the isobars rather than directly across them.

You may have noticed that the isobars are rather like the height contours on a normal map. This is a helpful way of thinking as we can visualize 'deep depressions' and 'ridges of high pressure'. In addition, the closeness of isobars can give us an idea of the 'steepness' of the pressure variation. **Where the lines are very close together, the pressure changes rapidly and so the winds are strong.** Where the isobars are far apart, the pressure changes are gradual and so the winds are light.

Activities

Where is the high pressure when the wind blows?

We have seen that the spin of our planet means that, in the northern hemisphere, the winds blow in an anticlockwise direction around depressions. We can

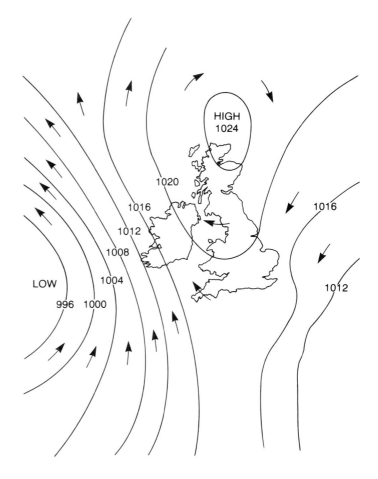

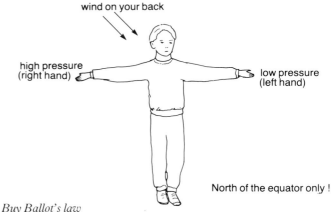

Buy Ballot's law

find where the high and low pressure areas are by this simple activity.

Stand with your back to the wind. Low pressure is to your left and high pressure is to your right.

How would this 'law' (known as Buy Ballot's law) need to be rewritten for use in Australia?

Weather maps

A report from a weather station tells us what the weather was like at that one place. To help us to predict accurately what the weather will be like, we need to put information from a large number of weather stations onto a map. The things we want to put on our map include: temperature; cloud cover; wind direction; wind strength; pressure; precipitation (rain, snow, and hail).

So that the map does not become too crowded, symbols are used.

Cloud

The symbol for cloud cover is a circle. Parts of the circle are then shaded in to show how much of the sky was covered by cloud.

Wind speed and direction

The direction of the wind is shown by a straight line attached to the cloud cover symbol. To understand the symbol you need to know the points of the compass.

The wind speed is shown by small lines added to the tail of the wind direction line. As the wind gets stronger, more lines are added.

The wind speed is given in knots. This is the same as the unit for speed used by seamen. 1 knot stands for 1 nautical mile per hour. This is just more than 1 mile per hour (1.85 km/h).

Precipitation and other weather conditions

Special symbols are used to show rain, snow, thunder storms and other types of weather. These symbols can be put together to give even better descriptions. For example, the snow symbol with the shower symbol means 'snow shower'.

Temperature and pressure

The pressure and temperature readings for a particular weather station are given in numbers by the side of the cloud cover symbol.

The symbol shown opposite tells us the following:

a) Three-quarters of the sky was covered with cloud.

b) The wind was blowing from the east.

c) The wind speed was between 8 and 12 knots.

d) The temperature was 9°C and the pressure was 996 mb.

e) It was raining.

CLOUD

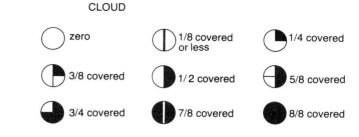

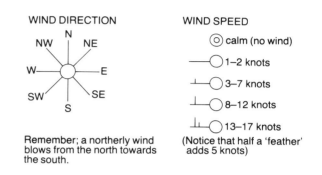

Remember; a northerly wind blows from the north towards the south.

(Notice that half a 'feather' adds 5 knots)

WEATHER

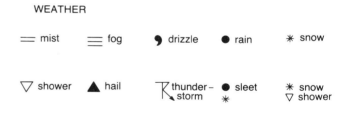

TEMPERATURE AND PRESSURE

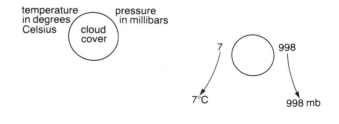

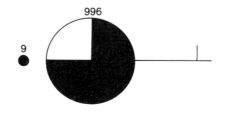

History of a typical depression

In the map opposite, two high pressure regions are affecting the British Isles. High A is making air from the cold, polar region to the north move down. High B is making air from the warmer southern regions move up. Where the two types of air meet is called a **front**.

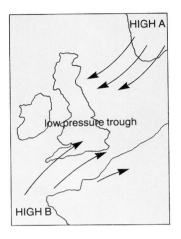

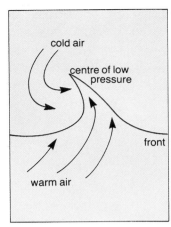

Start of a depression

Sometimes, far out in the Atlantic to the west of Britain, the front 'bends' to form a wave. As this wave moves a **depression** forms. As you can see from the diagram, the 'pointed' part of the depression contains warm air. On either side there is colder air.

Here are the isobars and wind directions of a well-formed depression. You will often see this pattern on television weather forecasts. Unfortunately it will bring rain and unsettled weather.

Cold air is denser than the warm air and so it moves under the warm air and pushes it upwards. This is what a depression looks like from the side. As the depression moves across the country the person at point X will notice that the weather changes.

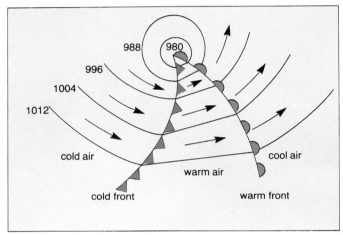

A depression as it appears on a weather map

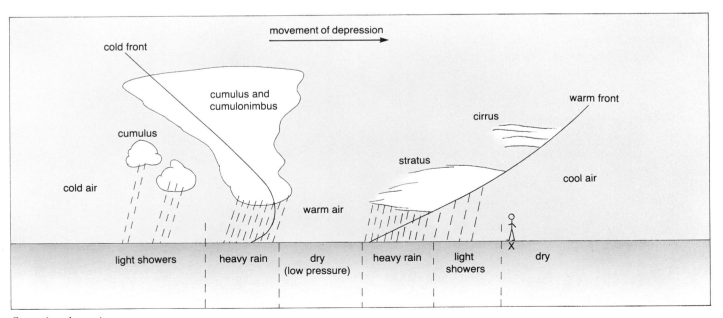

Stages in a depression

Changing the Earth's climate

The Earth's climate has not always been the same as it is today. For example there have been several ice ages when the average temperature was very low. This caused the polar ice caps to grow as the sea froze. The ice spread until it covered much of Britain.

The last ice age finished about 10 000 years ago. Since then the average temperature has been higher. No one knows exactly what caused the ice ages but most scientists think that it had to do with the Sun's activity or the orientation of the Earth.

Most of the things which can change our planet's climate, like volcanoes and the Sun's release of energy, are beyond our control. However, what humanity does can have a very serious effect on global weather patterns.

Volcanoes throw huge amounts of dust into the atmosphere.

Dust in the atmosphere

Dust is very small particles of solid matter. When volcanoes erupt huge clouds of ash are thrown high into the air. Some of this is so fine that it stays suspended in the atmosphere. This affects our climate because the dust particles reflect some of the energy radiated by the Sun back into space. Because less radiation reaches the Earth, the dust has a cooling effect. (Some scientists believe that if volcanoes become more active and throw enough dust into the atmosphere, it could cause another ice age.)

Human activities on Earth also cause dust. For example where large areas of forest have been cleared the wind can erode the soil and blow dust into the air. This also happens where large fields are ploughed by machinery.

In towns, factories and cars also cause dust. In fact some industrial cities produce a dome of dust which affects their climates. There are two main effects. Firstly much of the dust stays close to the ground (less than 10 km high). Here it reflects radiated heat from the town back down. This raises the temperature. Secondly, the tiny bits of dust help water vapour to condense. This increases rainfall.

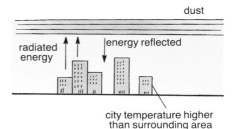
The effect of dust in the atmosphere

Nuclear winter hypothesis

Some scientists have suggested that in a large nuclear war, fires and explosions would throw huge clouds of dust and smoke into the upper atmosphere. The solid particles would stay suspended in the air and would reflect much of the Sun's radiation back into space. The Earth would be cold and dark for a very long time.

Some estimates suggest that several thousand megatons of nuclear warheads could cause a nuclear winter with a temperature of about −20°C. At these temperatures and in total darkness it is difficult to see how any life could survive.

No one has tested this idea by experiment; it is a **hypothesis** and not all scientists think that it is correct.

Could a nuclear war change our climate?

Carbon dioxide in the atmosphere

Air contains a small amount of carbon dioxide (about 0.03 per cent). Most of this comes from animal respiration and, of course, carbon dioxide is needed for plant photosynthesis (see page 126). However, the amount of carbon dioxide in our atmosphere affects our weather.

On page 206, we saw how shortwave radiation from the Sun is absorbed by the Earth. The warm Earth gives out radiation with longer wavelengths. Now carbon dioxide lets short wavelength radiation pass but absorbs some of the longer wavelengths. By 'trapping' this energy the carbon dioxide helps the Earth to heat up. This is sometimes called **the greenhouse effect**.

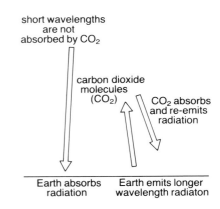

The greenhouse effect in the atmosphere

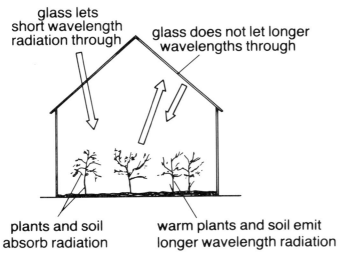

A real greenhouse

The greenhouse effect is good for us. Without it the Earth's average temperature would be about $-45\,°C$ instead of the present $+12\,°C$. However, in the last 150 years the amount of carbon dioxide in our atmosphere has risen by over 10 per cent. This is mainly because we have been burning fossil fuels such as coal and oil in our homes and factories. The large increase in carbon dioxide is thought to explain why the temperature of the northern hemisphere appears to be rising.

Scientists are worried that the rising temperature may melt the polar ice caps and make the level of the oceans rise. This could flood places like the Netherlands and parts of Britain.

We must study and measure what is happening in the atmosphere to make sure that the climate does not change too much. If it does, it could have very serious consequences.

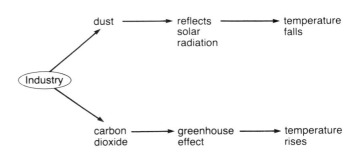

Industry affects our climate in different ways

Questions

1 In recent years there have been several large volcanic eruptions. Explain how this could lower the Earth's temperature.

2 What are the main causes of 'dust domes' over industrial cities?
What effects does the dust have on the weather near the city?

3 What would a 'nuclear winter' be like?

4 In some places, large forests are being cut down. This is called **deforestation**. Small trees and the branches of large trees are burnt. When the land is clear it is ploughed so that crops can be planted.
How does deforestation put more dust into the air?
How does deforestation put more carbon dioxide into the air?

Questions

1 A student tries to investigate whether shiny surfaces or dull, black surfaces absorb more energy. He sets up the apparatus below:

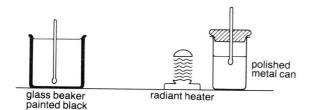

glass beaker painted black   radiant heater   polished metal can

a) The student has not designed a good experiment. Write down three 'mistakes' which he has made.
b) Draw a diagram showing how the investigation should be carried out.

2 Use the weather map below to answer these questions.

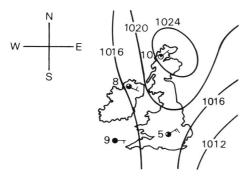

a) Is the pressure over Scotland high or low?
b) What is the temperature in London?
c) What is the wind direction near Land's End?
d) What is the cloud cover in Northern Ireland?

3 Zambia is a country in Africa, hundreds of kilometres from the sea. In the early morning the sky is usually very clear and the Sun shines. Later on small, white, fluffy clouds appear. These grow larger until very heavy rains fall.
a) Name the type of clouds described in the passage.
b) Suggest where the water which forms the clouds comes from.
c) Explain how the clouds form. In your answer use the words energy, evaporation, convection, condense (or condensation).
d) Explain why the pilots of small aeroplanes avoid flying through clouds like these.

4 The diagram below shows a depression.

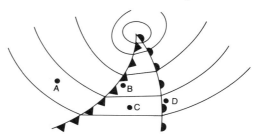

a) Draw the symbol for   i) a warm front   ii) a cold front.
b) What type of clouds would you expect to see at point A?
c) Where would you expect the heaviest rain to be?
d) Where would you expect it to be warmest?

5 A gardener has two maximum and minimum thermometers, one inside her greenhouse and one outside. She records the maximum and minimum temperatures and the weather conditions in a table. Some of her results are shown below.

Date	Outside		Greenhouse		Weather.
	Max.	Min.	Max.	Min.	
1st March	7°C	−3°C	15°C	2°C	Clear sky
2nd March	7°C	−2°C	13°C	2°C	Clear sky
3rd March	6°C	2°C	10°C	4°C	Cloudy at night

a) Explain why the gardener keeps delicate plants in the greenhouse during March.
b) Suggest why the maximum temperature inside the greenhouse is higher than that outside.
c) Suggest why the minimum temperature was much lower on March 1 than on March 3.

6 Dust in our atmosphere reflects some of the Sun's radiation away from the Earth. Volcanoes are one of the natural causes that put dust into the air. Human activities also create dust.
a) Explain how an increase in volcanic activity could cause another ice-age on Earth.
b) Describe one human activity which puts dust into the atmosphere.
c) Explain how some scientists think that a nuclear war would change our climate.

12 | Variation, inheritance, and evolution

What are cells?
What part do cells play in reproduction?
What is variation and why is it important?
How have plants and animals evolved?

The people in the photograph, and you, belong to one type or **species** of animal. Scientists call this species *Homo sapiens*. As you can see, the people have lots of things in common. For example, they have the same general body shape and their faces have similar features. However, even though they are all easily recognizable as humans there are lots of small differences between them. These **variations** are very important and have helped *Homo sapiens* to evolve over millions of years into very successful animals.

Activities

1   Cut out photographs of five famous people from newspapers or magazines.

List five things which help you recognize that they are all humans.

List five things which help you recognize that they are all different individuals.

2   Find two plants **of the same species**. These could be two trees, two garden plants or even two house plants but they must be of the same type.

List three differences. You could consider leaf shape, leaf size, leaf colour, bark pattern, flower shape, flower colour, height of plant, thickness of stem and other things.

Cells: the building blocks of life

What is a cell?

A cell is a unit of living material. **All** living things are made up of one or more cells.

The amoeba shown here is an animal made of only one cell. Plants and animals with just one cell are called **unicellular** organisms.

Multicellular organisms are made up of a number of cells. Human beings are multicellular and have millions of cells.

Plant and animal cells have similar jobs to do. They take in food, release energy, get rid of waste, grow and reproduce. Their structures, however, are not the same.

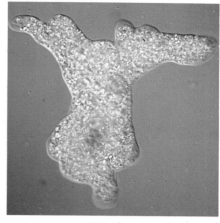

Amoeba

Cell structure

This is a photograph of some animal cells. These are cheek cells from the lining of someone's mouth. They have been stained to make them clearer to see.

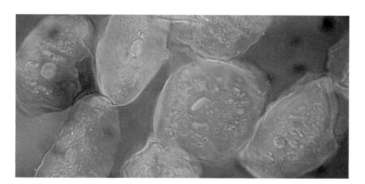

These plant cells come from the leaf of a pond weed:

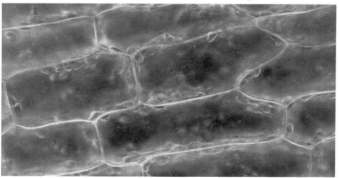

The main parts of the animal cell are shown in the diagram below.

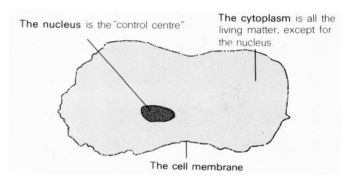

The **nucleus** is the "control centre"

The **cytoplasm** is all the living matter, except for the nucleus.

The cell membrane

The main parts of the plant cell are shown in this diagram.

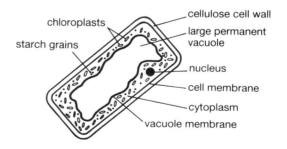

chloroplasts
starch grains
cellulose cell wall
large permanent vacuole
nucleus
cell membrane
cytoplasm
vacuole membrane

Questions

1 Why are cells sometimes called the 'building blocks of life'?
2 Why do you need a microscope to see a unicellular organism?
3 List **three** similarities between plant and animals cells.
List **three** differences between plant and animal cells.
4 Why were the cheek cells in the photograph above coloured with a dye?

From cells to organisms

Humans and other multicellular organisms are made up of different types of cell. Each type of cell or group of cells carries out a different job. As a result cells from a plant or animal may not all look the same. Some cells are designed:

. . . to carry messages . . .

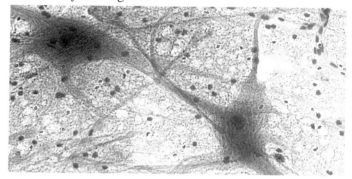

Nerve cells have long thin fibres to carry electrical impulses.

. . . to carry chemicals . . .

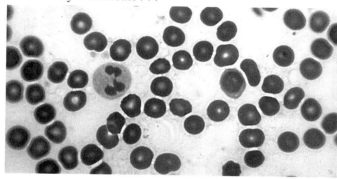

These red blood cells carry oxygen around the body.

. . . to absorb water . . .

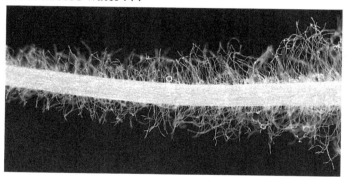

These cells from the root of a plant absorb water from the soil.

. . . to provide protection.

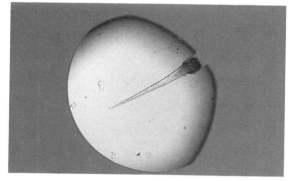

This is the sting cell of a nettle.

Specialized cells together make **tissues**. Muscle tissue is made of cells whose job it is to contract and cause body movement. Skin and bone are other examples of tissues.

Various tissues together make an **organ**. An organ performs important jobs in the body. The heart is the organ responsible for blood circulation.

Organs work together to form **organ systems**. The digestive system is an example; it is made up of various different organs such as the stomach and intestines.

An **organism** is a plant or animal which can exist on its own. It is made up of organ systems which work together to carry out all the functions of a living thing.

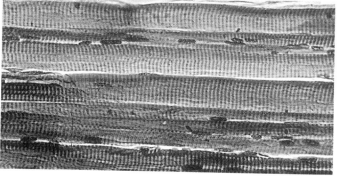

Muscle tissue

Questions

1 Name some of the organs in your body.

2 For one of these organs, list the tissues that make it up.

3 What is the difference between an organ and an organ system?

4 Why do you think cells are so specialized in complex animals and plants?

Sex cells and life cycles

Most animals and plants have cells in their body which are very different from any others. These cells are sex cells or **gametes**. Their job is to pass on information from one generation to the next.

Gametes are made in sex organs. In animals the male sex cells are **sperms** which are made in the **testes**. The female gametes are **eggs** and these come from the **ovaries**. Sperms are much smaller than eggs and, unlike eggs, are able to move on their own. They consist of a 'head' containing a large nucleus and a long tail which wiggles to propel the sperm along.

Egg cells are less complicated than sperms. A large nucleus is surrounded by cytoplasm which in some animals, particularly birds, contains yolk. The yolk provides food for the young animal as it grows in the egg. Human eggs have no yolk, a human baby obtains its food from its mother while in the womb.

Chapter 1 gives more information about gametes and sex organs.

In all animals, sexual reproduction involves the fusion of a sperm nucleus with an egg nucleus. The method by which sperms and egg are brought together, however, may be different.

Mating reptiles

In reptiles and birds fertilization is internal, eggs are laid and the young develop outside the mother's body.

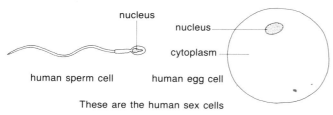

Human sex cells or gametes

The male frogs sits on the female's back. He produces sperm as she lays her eggs.

In humans and other mammals fertilization and development of the young takes place inside the female.

In most fish and amphibians fertilization and development of the young takes place outside the body of the female.

Life cycles

A life cycle describes the stages in development of members of a species from fertilization of one generation to fertilization of the next. The life cycle of humans and most other animals follows the pattern shown.

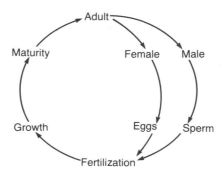

Life cycle of humans (and most other organisms)

Sexual reproduction in plants

Flowers are the reproductive structures of plants. They contain the reproductive organs which produce the male and female sex cells or gametes. Male gametes are contained in **pollen grains**, female gametes are **ovules**.

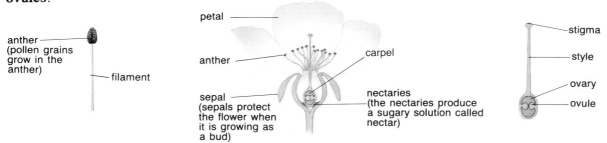

anther (pollen grains grow in the anther)

filament

petal

anther

carpel

sepal (sepals protect the flower when it is growing as a bud)

nectaries (the nectaries produce a sugary solution called nectar)

stigma

style

ovary

ovule

Insects such as bees visit flowers to collect nectar and pollen for food. The flower's shape and structure means that while they are doing this the hairy body of the bee becomes covered in pollen grains. As the bee visits other flowers, pollen is transported from the anther of one flower to the stigma of another. This is **cross pollination**. Sometimes pollen from a flower is deposited on its own stigma. This is **self pollination**.

Fertilization in plants

Fertilization, as in animals, involves the joining of the nuclei of the male and female gametes. It happens like this . . .

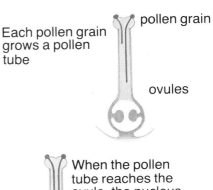

Each pollen grain grows a pollen tube

pollen grain

ovules

When the pollen tube reaches the ovule, the nucleus from the male cell travels down to join with the nucleus in the female cell

The pollen grains of some flowers are carried not by insects but by the wind. Instead of being sticky to enable them to become attached to the body of an insect, the grains are smooth and very light so that they can be blown over long distances by air currents. Grasses are wind-pollinated.

Anthers and stigmas hang outside the flowers to help release and capture pollen. Wind-pollinated flowers are usually much smaller than insect-pollinated flowers and are not so attractive to look at. These hazel catkins are a good example.

Activities

Take a flower from an insect-pollinated plant. Use tweezers and small scissors to separate the parts of the flower. Count the number of petals, anthers and carpels. Cut an ovary open and try to find the ovules inside.

Hazel catkins; wind pollinated flowers

Seeds and germination

After fertilization most of the flower dies, only the ovary survives. Inside the ovary the fertilized ovules grow and develop into seeds. A **seed** consists of a tiny embryo plant enclosed with a supply of food in a tough protective coat or **testa**.

As seeds grow the ovary grows as well. The fully developed ovary with its seeds inside is called a fruit. Bean pods, sycamore 'helicopters', dandelion 'parachutes' and plums are good examples of fruits. Their job is to help scatter or **disperse** the seeds to new areas so that the new plants won't be overcrowded when they grow or **germinate**.

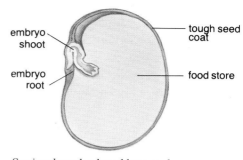

Section through a broad bean seed

dandelion fruit

'parachutes' and 'wings' catch the wind

Some seeds are scattered by the wind

sycamore fruit

burdock fruit

hooks catch on animals' fur

cherry

seeds are left behind in animal droppings

Some seeds are scattered by animals

sweet pea

On a hot day the dry pod bursts open

Some seeds are scattered by explosions

Germination

Seeds need water, oxygen and a suitable temperature before they begin to germinate. That is why many plants don't begin to grow until the spring when it gets warmer and the soil is neither frozen nor hard and dry.

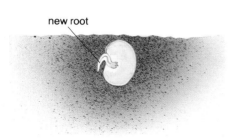

new root

*Water is absorbed and the seed swells. The testa softens and the young root (**radicle**) grows down into the soil.*

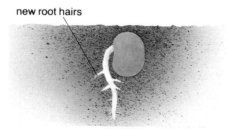

new root hairs

Root hairs develop to enable more water and dissolved mineral salts to be taken into the seed. Water helps prepare food reserves for use by the growing plant.

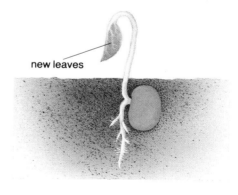

new leaves

*The young shoot (**plumule**) grows upwards through the soil, develops leaves and soon begins to make new food by photosynthesis.*

Questions

1 Why are seeds dispersed away from the parent plant?

2 Explain how the fruit of the sycamore is designed to carry seeds away from the parent plant.

3 Explain why fruits like cherries are brightly coloured and sweet tasting on the outside and why the seed inside has a hard coat.

4 Design an experiment to show that cress seeds need water and warmth before they will germinate.

Asexual reproduction

Not all organisms have gametes (sex cells). They reproduce **asexually**.
This type of reproduction needs only one parent. This means that the
offspring will be identical to each other and to their parent. These identical
offspring are called **clones**. Asexual reproduction is usually much quicker
than sexual reproduction. It enables simple plants and animals to
reproduce quickly and efficiently.

There are a number of methods.

Fission

Very simple organisms like amoeba can reproduce by
dividing into two equal parts. These grow to full size
and the process is repeated. Bacteria can achieve rapid
rates of fission; once every 20 minutes under ideal
conditions.

Asexual reproduction in amoeba is by 'splitting'

Budding

This process differs from fission in that the parts
produced are not equal. In yeast, a swelling called a
bud appears on the outside of the cell wall. The bud
grows and then breaks free. Sometimes buds remain
attached to the parent cell and long chains of cells
develop.

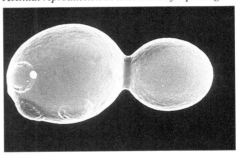

Budding in yeast

Spore production

Fungi like pin mould produce vast numbers of spores
inside spore cases held up in the air on stalks. Spores
are very light and have a protective coat around them.
Because they are very small and light they can be
carried great distances by air currents.

Pin mould with spore cases

Activities

Dampen a thin piece of white bread with water and put it into a polythene
bag. Seal the bag tightly. Put a label on the bag and place it in a warm
place – an airing cupboard, for instance. After a few day examine the
bread carefully. You should be able to see a healthy growth of mould.

The 'furry' appearance is produced by lots of vertical stalks growing from
the surface of the bread. If you have a hand lens you may be able to make
out the mass of thin branching fibres spreading through the bread.

1   Why do you think this mould is called pin mould?

2   Where do you think the mould came from?

Asexual reproduction: propagation

Some plants, as well as being able to reproduce sexually by producing seeds, are also able to reproduce by growing new parts which can live as separate plants. This is a form of **asexual** reproduction since no gametes are involved.

Strawberries send out special stems called **runners** that spread over the ground. At the tips of these stems a bud forms which grows into a new plant.

Strawberry plants

Rhizomes are underground stems. Couch grass has a rhizome. Notice how roots and leaves grow at intervals along it.

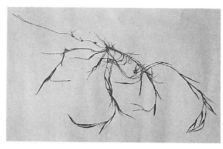

Couch grass

Tubers are also underground stems. Potatoes are an example. Although swollen with food, the potato, like any stem, has buds – the 'eyes'.

Potato tuber

Bulbs are large underground buds with swollen leaves full of food.

Onion bulb

Corms are often mistaken for bulbs but instead of being a bundle of swollen leaves they are a flat, swollen stem.

Crocus corm

Humans have made use of the plant's ability to reproduce asexually. Stocks of identical plants can be built up by taking **cuttings**. Many of our well-known house plants like geraniums are produced from stem cuttings. Others such as the African violet are grown from leaf cuttings. Taking cuttings is really quite simple, all you do is place a piece of cut stem or a leaf into some moist compost and leave it to develop its own roots. Some are much easier than others. The new plants must be looked after very carefully.

Fruit and rose growers use the technique of **grafting**. A V-shaped cut is made into the stem of a healthy plant (the **stock**). Then a stem from another plant of the same type (the **scion**) is trimmed and fitted into the cut. The two parts are then tightly bound together and sealed. After a few weeks the two parts will have grown together. Grafting is useful if lots of a popular fruit or flower are needed for sale.

Activities

Taking cuttings

Carefully cut a piece of stem about 10 cm long from a geranium, 'busy lizzie' or other house plant. (Ask permission first!) Take off the bottom leaves leaving three or four at the top. Place your cutting in a jar so that the bottom 4 cm is in water. Leave it until you can see roots growing from the stem. Once these new roots are about 2 cm long you can plant your cutting in some moist compost.

Because this is asexual reproduction, your new plant will be just like its parent.

Variation

Every species has characteristic features of shape and structure by which it can be recognized and identified. Look at your classmates and you should recognize that they have similar features to you; features that make them human beings.

Variation in humans

However, you should also be aware that individual students in your class are not exactly the same as you or each other. Different hair colour, height, weight and skin colour are examples of differences that we call **variation**.

Characteristics that are distinct, such as being able to roll your tongue or having ear lobes are examples of **discontinuous variation**. You either have the characteristic or you don't.

Are you a roller or a non-roller?

Other characteristics are not so easily separated, such as the difference in human body weight. If you measure and record the height of your classmates, you should find that they form a complete range. This is an example of **continuous variation**.

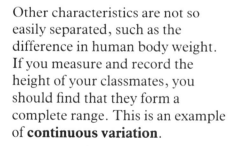

Height is continuously variable.

Do you have ear lobes or not?

Changes in the environment such as different amounts of food, light and space can have an effect upon the way individuals develop. Height and weight can be influenced by what sort of food we eat and how much we eat. Our skin colour is dependent on how much sunlight we are exposed to. A typical European on holiday in a sunny country will change colour for a short while!

This sort of variation is often referred to as **acquired variation**.

Activities

Make a family tree of your relatives and try to find out about their characteristics. Ask about things like hair colour, eye colour, height, colour blindness, tongue rolling ability etc.

Can you see where your characteristics came from?

Passing on information

Inside the nucleus of each one of your cells are structures called **chromosomes**. These are fine threads of material which carry 'bits' of information about what you look like. These bits of information are called **genes** and each one controls one or more characteristics that you have inherited from your parents.

Human cells have 46 chromosomes altogether. These are paired off, making 23 pairs. Each chromosome in a pair is like its partner and carries information about the same feature although this information may not always be the same. For example, one of a pair of chromosomes with information about hair colour may carry the instruction 'be blond' while the other may carry the instruction 'be dark brown'. These pairs are called **homologous pairs** of chromosomes.

We inherit one chromosome in a homologous pair from our mother and the other from our father.

When gametes are made the number of chromosomes must be halved. Homologous pairs are separated inside the sex organs so that the sperms and eggs, pollen and ovules only possess one of each pair. At fertilization the two 'half sets' are put together to make a full set of pairs in the new individual.

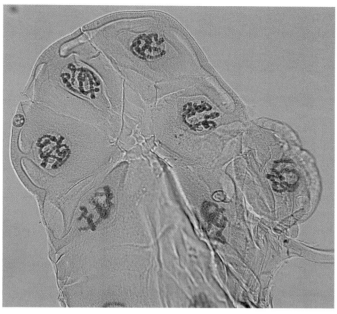

Cell showing chromosomes dividing

As a new baby or seed grows, more cells must be made, each one having its own set of chromosomes carrying just the same information as the original fertilized cell. The stages of cell division are shown in the diagram below. (Only four chromosomes are shown to make things simpler.)

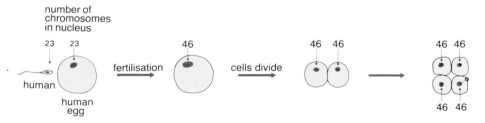

Fertilization and cell division

Questions

1   What is the difference between a chromosome and a gene?
2   A mouse has 40 chromosomes in each of its cells.
   a) How many homologous pairs does it have?
   b) How many chromosomes are there in the sperm of a male mouse?
3   A pea plant has 14 pairs of chromosomes in each of its cells.
   a) How many chromosomes will there be in a pollen grain cell?
   b) How many chromosomes will there be in an ovule?
   c) Explain why a pea plant grown from a seed will have 28 chromosomes in each of its cells.

Each chromosome makes a copy of itself.

The copies separate. One chromosome from each pair goes to opposite sides of the cell.

The cell divides: new nuclei form. Each daughter cell has the same chromosomes as the parent.

Patterns of inheritance

Many of our features, for example hair colour and eye colour, are controlled by a pair of genes. The colours of your hair and eyes were determined at fertilization when the nucleus of your father's sperm fused with the nucleus of your mother's egg. The passing on of characteristics from one generation to the next is called **heredity**. The study of heredity is called **genetics**.

Let us think about hair colour. At fertilization the two genes for hair colour which come together could carry the same instructions, for example 'be blond'. In this case the person will have blond hair. When the two genes are the same we call the person **homozygous** for the character.

Of course the two genes could carry different instructions, for example 'be blond' and 'be dark brown'. In this case, one of the genes will usually be **dominant**. This means that one gene will dominate the other and will control the character. The gene which is not dominant is called **recessive**. Now 'dark brown' is dominant over 'blond' so the person will have dark hair. When two genes are different we call the person **heterozygous** for the character.

If both parents are heterozygous for a particular character then their children could inherit genes in a number of different combinations. A useful way of showing the possible gene combinations is to use a Punnett square (named after the famous geneticist Reginald Punnett).

The Punnett square drawn here shows the possible outcomes of a cross between two parents heterozygous for eye colour. B represents the gene for brown eyes and b the gene for blue eyes. The gene for brown eyes is dominant over the gene for blue eyes – this is why it has a capital letter!

When we look at a person we cannot always tell what sort of genes they have. For example a brown-eyed person could have two genes for brown eyes (homozygous) or one gene for brown eyes and one gene for blue eyes (heterozygous). To avoid confusion we must use two more terms: **phenotype** – the outward appearance of the person, and **genotype** – the kind of genes the person has.

Hair colour is controlled by genes. Each of the blond girl's cells contain two genes for blond hair.

Male parent heterozygous brown (Bb)

		Sperm	
		B	b
Female parent heterozygous brown (Bb) — Eggs	B	BB	Bb heterozygous brown
	b	Bb heterozygous brown	bb

B = brown eyes   b = blue eyes

A Punnett square

Questions

1 **a)** What does *homozygous* mean?
   b) If an egg carrying a gene for blue eyes is fertilized by a sperm carrying a gene for brown eyes, will the person be homozygous or heterozygous for eye colour?

2 **a)** Write down your *phenotype* for   i) hair colour   ii) eye colour.
   b) Write down the *genotype* for a blue-eyed person.
   c) Why couldn't you be sure about the genotype of a brown-eyed person?
   (*Hint:* Use the words dominant and recessive in your answer.)

3 Draw a Punnett square to show the possible outcomes of a cross between a brown-haired woman and a blond man. (Use the letter B for the brown hair gene which is dominant and b for the recessive blond gene.)

4 A person with brown hair has it dyed red. Have they changed their genotype? Explain your answer.

Mendel and inheritance

Gregor Mendel, an Austrian monk, published the results of his experiments on inheritance in 1865. No one bothered with his work at the time and it was not until 1900 that his work was really recognized as the basis for all modern genetics.

Mendel carried out simple experiments using garden peas. He noticed a number of differences in their characteristics. For example, some varieties have short stems and others have long stems, some have smooth seeds and others have wrinkled seeds.

During his experiments, Mendel carefully transferred pollen from the anther of pure-breeding tall plants to the stigma of pure-breeding short plants. He collected seeds from the plants, planted them, and carefully observed the growth of the new **hybrid** plants. He found that they were all tall.

Clearly the gene for tallness was dominant over the gene for shortness and the offspring were therefore all tall.

Mendel allowed these hybrid plants to flower and pollinate themselves. When seeds were produced he planted them and carefully recorded the results of the cross.

He found that about three quarters of the offspring were tall and one quarter were short. We say that the ratio of tall plants to short plants was 'three to one' (3:1) because there were three tall plants for each short plant.

Tall and dwarf pea plants

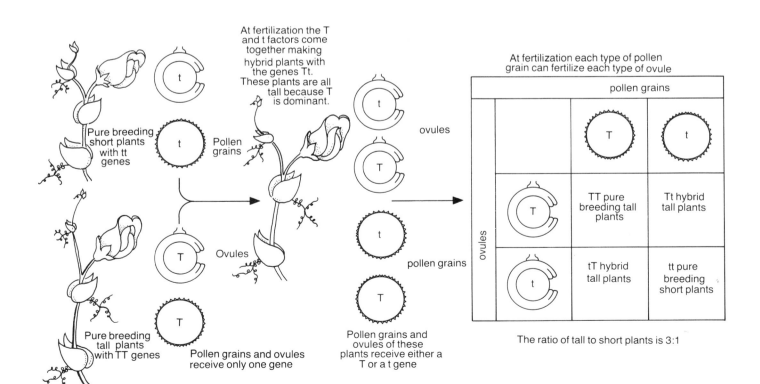

The ratio of tall to short plants is 3:1

Boy or girl?

Genetics is involved in determining what sex we are. Chromosomes that determine whether we are male or female are called **sex chromosomes.** There are two types, one is called an X chromosome and the other, shorter one is called a Y chromosome. Females have two X chromosomes, males one X chromosome and one Y chromosome.

At gamete formation all the eggs from the mother will carry X chromosomes. However, only half the sperms of the father will carry an X chromosome, the other half will carry a Y chromosome.

If a sperm carrying an X chromosome fertilizes an egg the child will be female. On the other hand if the sperm contains a Y chromosome the fertilization will result in a male child.

It is the sperm therefore which determines the sex of children. There is a 50 per cent chance of a boy and a 50 per cent chance of a girl. This is why the human population, like many others, is roughly half male and half female.

Sex-linked inheritance

Genes that are present on the X chromosome are more likely to affect a male than a female. This is because the Y chromosome is shorter and therefore some gene positions will be missing. The gene for colour blindness is recessive and is on the X chromosome in a position which is unfortunately not present on the Y chromosome. As a result the message on the gene cannot be dominated and the male will be colour blind. In females, colour blindness is rare. A woman will only be colour blind if both X chromosomes carry the same recessive gene.

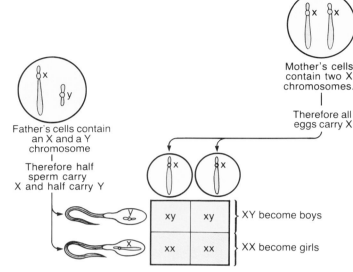

Sex determination in humans

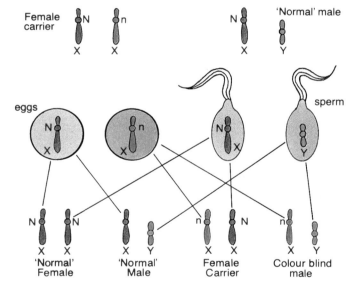

Genetics and colour blindness

Questions

1 Write down the phenotype (male or female) and the genotype (XX or XY) for **a)** yourself **b)** your teacher **c)** your best friend.

2 It is possible to find whether an embryo in the womb of a pregnant woman has XX or XY chromosomes.

    a) Explain how this tells whether the baby will be a boy or a girl.

    b) Suggest one reason why this information may be important to a parent.

    c) Do you think parents should be able to find out the sex of their baby before it is born? Explain your answer.

3 Write down *five* problems encountered by colour blind people.

Mutations

Sometimes when cells divide, the structure of a chromosome may become altered or a gene copied incorrectly. These changes are called **mutations**. All descendants of an individual having a mutation are called **mutants**.

Chromosome mutations

When gametes are formed in the sex organs there is a chance that changes in the number or structure of chromosomes can take place. Some chromosomes break in two places and the piece in the middle falls away taking its genes with it. This will seriously affect the development of an organism.

On other occasions, homologous pairs of chromosomes do not separate properly and both go into the same gamete. This kind of chromosome mutation is the cause of **Down's syndrome**. People having this condition possess three 'twenty-first' chromosomes instead of two. This possession of an extra chromosome causes physical and mental problems for the sufferer.

A chromosome mutation can cause a woman to produce an egg with 24 chromosomes instead of 23. If the egg is fertilized the baby develops Down's syndrome.

Gene mutations

A chemical change which alters the message carried by a single gene is called a gene mutation. Gene mutations that occur in gamete-producing cells are transmitted to all the cells of the offspring and may therefore affect the future of the species.

Haemophilia is a disease that prevents blood clotting. It is sometimes called the 'bleeding disease'. About fifty out of every million people become haemophiliacs. Queen Victoria was a carrier of the disease. She was born with one mutant gene and passed it on through the royal families of Europe.

The formation of 'new' genes by mutation has probably had a significant effect upon the range of plant and animal species that live on Earth today.

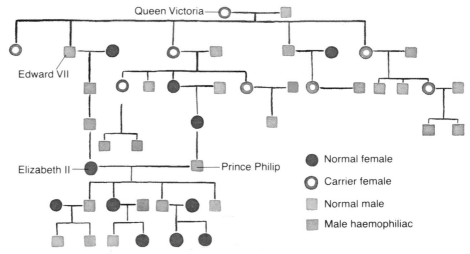

Haemophilia in the Royal Families of Europe: can you identify the unnamed members?

Questions

1 Explain the difference between a chromosome mutation and a gene mutation.

2 What are mutants? Name a mutant condition that you have read about.

3 Why wasn't Queen Victoria a haemophiliac even though she carried the mutant gene for haemophilia?

4 Warts, moles and other skin blemishes are caused by mutations of skin cells. These features are not passed on from one generation to the next, they die when the person dies. Why is it, therefore, that haemophilia is inheritable?

What is a species?

A **species** is a population of organisms which can breed together and produce fertile offspring. Usually individuals of different species will not interbreed. If however breeding does take place then the offspring are likely to be sterile. This is because the species will have different numbers of chromosomes in their cells.

The horse and the donkey are different species but because they are very similar they can interbreed. The offspring of the horse-donkey cross is called a **mule**. A mule is sterile and cannot produce baby mules. The only way to get mules is to cross a horse with a donkey.

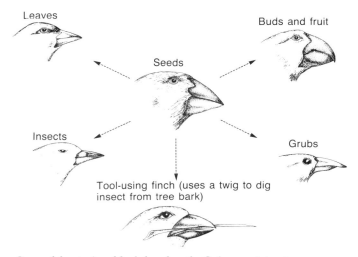

A mule: healthy but sterile

How are new species formed?

Before a new species can be formed it is necessary for a large group of animals or plants to be split up into separate groups. These groups must be **isolated** from each other so that there is no possibility of genes being exchanged between them. Breeding will continue within the groups producing more generations. Over thousands of years, mutations and variations will produce organisms which differ so much from the original group that they form a new species.

Geographical isolation

Populations of animals or plants may become split up by natural barriers like mountain ranges, rivers and seas. The finches on the Galapagos Islands in the Pacific Ocean are a good example. Mainland finches are all the same: they have short, straight beaks for crushing seeds. On the Galapagos Islands there are thirteen species, differing mainly in the shape of their beak. It would seem that the original population of finches was split up and isolated on the separate islands. Each smaller group adapted to feeding on the different types of food found there.

Leaves

Buds and fruit

Seeds

Insects

Grubs

Tool-using finch (uses a twig to dig insect from tree bark)

Some of the species of finch found on the Galapagos Islands

Behavioural isolation

Courtship before mating is important in the behaviour of many animals. It helps members of the same species to identify one another and to choose a healthy mate. Birds use visual and sound signals. Among animals, smell is an important factor in recognition.

Individuals that are not able to produce the appropriate signal or not produce the signal at the right time are not going to attract a mate.

A bird that sings the wrong song or has the wrong colour feathers will not be able to breed.

A peacock displaying to attract a mate.

Evolution: the theory

The theory of evolution offers an explanation for the production of new species from earlier, simple life forms over millions of years.

In his book 'On the Origin of Species', Charles Darwin explained his theory of how evolution could come about by a process of **natural selection**. He received support from another naturalist, Alfred Russel Wallace, who had come to the same conclusions as Darwin at about the same time.

Darwin had investigated the plants and animals in different parts of the world when he was on board a survey ship, *HMS Beagle*. He was particularly impressed by the wildlife on the Galapagos Islands in the Pacific Ocean.

Despite some strong opposition following the publication of his book in 1859, Darwin's theory has stood the test of time and is today widely accepted.

Charles Darwin (1809–1882)

The theory of natural selection

The Darwin/Wallace theory of natural selection can be summarized as follows:

1 Within any population of living things there is variation.

2 Even though all species produce large numbers of offspring, natural populations stay fairly constant.

3 There must be a struggle for survival within populations.

4 Some individuals are better adapted to their surroundings. These are more likely to grow and reproduce passing on their advantage. Others without it, will die. This is sometimes referred to as '**survival of the fittest**'.

So, particular organisms have been naturally selected from their population because they are better adapted to the environment than their fellows.

Alfred Russel Wallace (1823–1913)

Natural selection in action

There are a number of examples of natural selection that have been studied during the past 50 years.

Cepaea nemoralis is the common land snail found in woods, fields, hedges, sand dunes and rough ground all over Europe. The shells of this species show considerable variation. The background colour may be various shades of brown, pink or yellow and the shell may have up to five dark bands around it. Both colour and banding is **inherited**. Snails which do not blend in with their surroundings are easily spotted by birds who will eat them. In woods with only a little undergrowth, banded snails are at a disadvantage and will be selected out, but in undergrowth and on rough ground banded snails are well camouflaged and are so be more likely to survive. Snails with unbanded shells will be easily seen by birds and are therefore likely to be eaten!

Which of these snails will be selected out?

Artificial selection

Since civilization began, thousands of years ago, humans have been selectively breeding domesticated animals and plant crops. Darwin used evidence from this **artificial selection** to account for evolution by natural selection.

New varieties of animals and plants are produced by deliberately selecting those individuals that have desired characteristics and breeding from them.

All dogs, even breeds as different as the Great Dane and Chihuahua, belong to the same species and have descended from the wolf.

Pug

Wolf

Irish wolfhound

Chiahuahua

Great Dane

The danger of inbreeding

Continued breeding of closely related individuals is common, especially in the production of 'show' animals like cats and dogs. Unfortunately this practice can lead to infertility and reduced resistance to disease. This is why it is important to investigate the family tree or **pedigree** of an animal before buying it.

Questions

1 Explain the difference between *natural* selection and *artificial* selection.

2 In what ways do humans struggle to survive? (*Hint:* it may help to think of people in developing countries.)

3 A modern cow produces several gallons of milk each day. Suggest how farmers have artifically selected cows to give so much milk.

4 Champion male race horses are often 'put to stud'. Race horse owners then pay thousands of pounds to have their female horses mated with the ex-champion. Explain why.

5 What do you think you should look for in a pedigree before buying a dog?

Artificial selection breeds faster racehorses.

Evolution: the evidence (1)

Evidence from fossils

Fossils are any sort of remains of a once-living organism preserved in the rocks of the Earth's crust.

Under special conditions the entire body of an organism may be preserved after it dies. Insects trapped in the sticky sap of ancient trees can now be observed embedded in amber. In the early part of the twentieth century frozen woolly mammoths were found in Siberia. Their meat was still edible after thousands of years in the deep freeze!

Total preservation of organisms is rare; usually the soft body parts decompose leaving only the bones or shells. When these become covered with sediment, sand or gravel, they produce fossils a few thousand years later when the sediment becomes rock.

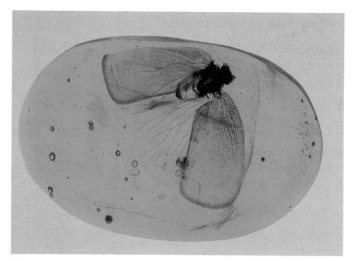

Fossilized insects

The Grand Canyon – 2000 million years of history

The Grand Canyon in Arizona is the deepest crack in the Earth's surface, about a mile deep.

The rocks of the canyon contain thousands of fossils. Each layer of rock provides evidence of the order in which animals and plants appeared on Earth over the past 2000 million years.

Near the top, fossils of reptiles such as dinosaurs can be found. Further down, the number and complexity of fossil species decreases. Halfway down, the fossils of fish first appear and below three quarters of the way down there are no signs of life at all.

Grand Canyon, USA

Activities

Making 'fossils'

1 Put some soft, moist sand into a small dish.

2 Push an object such as a shell or a bone into the sand so that a clean impression is made.

3 Make a cardboard frame around the impression making sure it is both carefully sealed and pushed well into the sand.

4 Pour Plaster of Paris into your mould and leave to it set.

5 Carefully remove the plaster cast from the mould and you have a 'fossil'.

Of course, this is not a *real* fossil but the activity shows how some fossils may have formed millions of years ago.

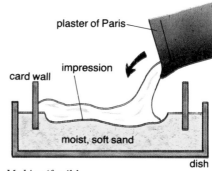

Making 'fossils'

Evolution: the evidence (2)

Evidence from anatomy

If you compare the skeleton of one mammal with another it is easy to spot that some parts of the body have a similar structure. Notice in the diagram how the limbs of various mammals share the same basic structure even though they have completely different functions and are different shapes.

This provides evidence that animals like these evolved from a common ancestor.

Other similarities can be recognized in the bodies of different species of mammals. For example the blood circulatory system and the digestive system are very similar in each type of animal.

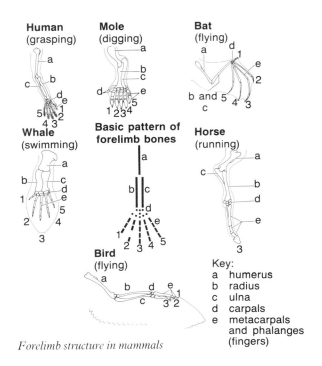

Forelimb structure in mammals

Key:
a humerus
b radius
c ulna
d carpals
e metacarpals and phalanges (fingers)

Evidence from geographical distribution

The distribution of animals on the oceanic islands provides strong support for the theory of evolution.

An oceanic island is one which has never been connected to the mainland. Usually they are formed by the eruption of volcanoes lying deep beneath the sea. As molten rock from the volcano cools down it solidifies and forms an isolated home for any organism that can get there from the mainland.

Study of the life on oceanic islands shows that many of the species on them had relatives on the mainland but over thousands of years they have evolved so that they are now very different.

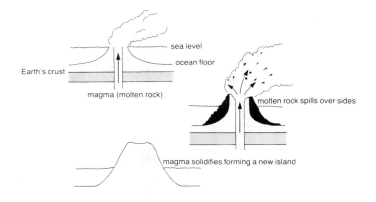

Formation of oceanic islands

Questions

1 The 'limbs' of the bat and the whale are called **homologous** because they have the same basic structure.

    a) Look at the diagrams on this page and list *two* similarities between them.

    b) Why does the similarity between the bones of the bat and the whale suggest that they have evolved over millions of years from the same type of animal?

2 Hawaii is a volcanic island.

    a) Explain how islands like Hawaii form.

    b) Suggest two ways that animals could have reached Hawaii from the mainland.

    c) Suggest two ways that plants or their seeds could have reached Hawaii.

Evolution of humans (1)

There are more humans on Earth than any other large animal. They live in the hottest deserts as well as the coldest parts of the planet. People have climbed the tallest mountains, explored the deepest seas and even travelled to the Moon. In fact our species, *Homo sapiens*, has evolved into the most successful species on Earth. This is not because we are the strongest or the quickest but because we are the most intelligent. We can plan, communicate in speech and writing, and can make and use tools. We need to ask 'what did we evolve from?'

Where did the human race come from?

Our ancestors are not alive today. There is some convincing evidence to support the view that humans and the apes of today, the gorilla, orang-utan and chimpanzee, share a common ape-like ancestor that lived 50 million years ago.

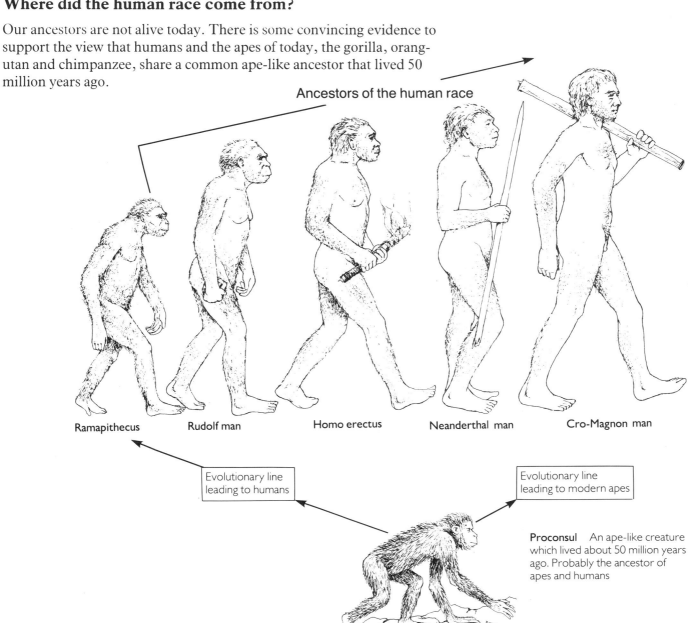

Ancestors of the human race

Ramapithecus   Rudolf man   Homo erectus   Neanderthal man   Cro-Magnon man

Evolutionary line leading to humans

Evolutionary line leading to modern apes

Proconsul   An ape-like creature which lived about 50 million years ago. Probably the ancestor of apes and humans

Of course a lot of guesswork has gone into showing what our early ancestors looked like. However, from skeletons and other fossils, scientists and artists have been able to produce the drawings shown above.

Evolution of humans (2)

We will never know why a group of ape-like creatures moved down from the trees onto the grassland millions of years ago. One possible explanation could be that a small group was driven out of the forests by the rest because of competition for food and space.

On the grasslands our ancestors had no protection and so environmental pressures selected only those individuals who were better adapted to their new surroundings. Over millions of years a new kind of animal, *Homo sapiens*, evolved.

Being able to think quickly was a big advantage, those that couldn't soon starved or got killed by animals. Natural selection allowed a large brain case and brain to evolve.

early ancestor

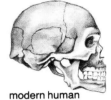

modern human

Notice how a large space for the brain has evolved

Feet slowly changed from gripping structures, necessary in trees, to more rigid body parts suitable for heel and toe walking and running.

In humans the pelvis is held directly above the legs. This provides a more upright posture and more efficient movement. Notice how your knees and feet are close together when you stand upright. Unfortunately there are some disadvantages to this arrangement of bones. A large number of people today suffer from back problems such as slipped discs and some women have difficulty in giving birth because of the relatively narrow pelvic girdle.

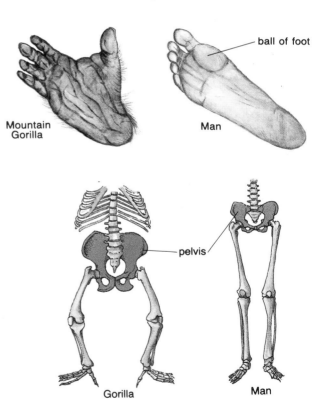

The human pelvis is much shorter. This allows the knees and feet to be close together when walking or running.

Questions

1 Why is it wrong to say that humans have descended from apes? Write a better statement about the link between humans and apes.

2 Why was it an advantage for our early ancestors to stand upright when they moved from the forests to the grasslands?

3 Suggest how Zulus and other African natives are adapted to survive in hot climates.

4 Explain how our large brains have enabled us to dominate all other large animals. (*Hint:* think about hunting, farming, building etc.)

5 What sort of inventions have made humans successful?

1 The diagram below shows a cell.

a) Did the cell shown come from a plant or an animal?
Give two reasons for your answer.
b) Each plant and animal cell has a 'nucleus'.
What does the nucleus do?
c) An amoeba is an animal made up of a single cell. Draw an amoeba. Label the main parts.

2 Sex cells are called *gametes*. Gametes contain *chromosomes* which pass on information from one generation to the next. In animals, the male gamete is called a *sperm*.

    a) Name the gamete made by female animals.
    b) This diagram represents a sperm.

      i) What is the tail for?
      ii) Where are the chromosomes?
c) The sex of humans is controlled by X and Y chromosomes. Women have XX chromosomes and men have XY chromosomes.
      i) What type or types of chromosomes could you find in a sperm?
      ii) What type or types of chromosomes could you find in a female gamete?

3 **a)** Some fruit trees will only give fruit if their flowers are pollinated by pollen from a different variety of tree.
      i) Explain what is meant by pollination.
      ii) Explain how pollen can reach one fruit tree from another.
    b) Some fruit trees can be propagated by taking cuttings.
      i) What is meant by 'propagation by cuttings'?
      ii) Why is a tree which grows from a cutting exactly like its parent?

4 The photographs below show a cart horse, used for pulling heavy loads, and a racehorse, used for sprinting.

Both horses are varieties of the same species. They have been bred using artificial selection to enable them to do their different jobs.
    a) How is the cart horse adapted to do its job?
    b) How do you think breeders have used artificial selection to produce the cart horse?
    c) Modern racehorses have been produced by breeding from mares (females) and stallions (males) that can run fast. Give two reasons why a person should check the pedigree of a racehorse before buying it.
    d) How would you prove that racehorses and cart horses are varieties of the same species?

5 A homozygous blond man has a heterozygous brown-haired wife. Show that there is a fifty/fifty chance that their first child will have blond hair. Use the symbols B for brown and b for blond. (Brown is dominant.)

6 Humans have evolved into the most successful of all the mammals. Their large brain lets them plan, make tools and machines, and communicate through language.
    a) How does 'thinking' help eskimos survive in very cold conditions?
    b) How have machines helped to improve our food supply?
    c) If you visit a foreign country but do not know the language it is difficult to make yourself understood or understand what people are saying. Suggest how you could communicate with people in such circumstances.

Ideas for investigations and extension work

Chapter 1

1 Design an investigation to test the following hypothesis:
'The strength of the human forearm is directly related to body mass.'

2 Design and carry out an experiment to test the effect of a cola drink on teeth.

3 'Twice as much saliva will double the speed of starch digestion.' How would you test this idea?

4 Design and carry out an experiment to compare the reducing sugar content of different kinds of sweets.

5 Does so called 'junk food' really deserve its name?

6 How nutritional are school meals?

7 Carry out a survey of the eating habits of school students. Suggest ways of informing teenagers about the importance of healthy eating.

8 Design and carry out an investigation to compare the structure and pH of a selection of toothpastes. Are smokers' toothpastes different from the others?

9 Design a piece of apparatus that would enable someone to compare the amount of tar present in the smoke from a selection of cigarettes. How would you use the information gained by using this apparatus to dissuade school students from smoking?

10 Design an investigation to see if:
   i) hearing is the same in both ears
   ii) sight is the same in both eyes.

11 'Bacteria are everywhere'. Design and carry out an experiment to test this hypothesis.

Chapter 2

1 Design an experiment to show how the speed of a runner varies during a 200 m race.

2 How could you test the following statement? 'On many roads people drive too close to the car in front of them to stop safely.'

3 The planet Earth travels around the Sun in a roughly circular orbit of radius 150 000 000 km. One orbit takes about 365.25 days to complete. Calculate the average orbital speed of the Earth in km/s. (Circumference of a circle is given approximately by $44/7 \times$ radius.) Put your answer in a table with the headings 'name of planet', 'average distance from the Sun', 'length of year in Earth days', 'average speed'. Fill in the table for the other planets in our solar system. Do you see any patterns?

4 Design an experiment to measure the acceleration of a bicycle rider from rest.

5 a) Design an experiment to measure the maximum frictional force (limiting friction) between a metal block and a plank of wood.
   b) 'Talcum powder, water, and oil are good lubricants.' How would you test this statement?

6 'Drinking alcohol increases a person's reaction time. Some people are affected in this way more than others.' Design a series of experiments which could be used to test this statement.

7 Discuss the following idea: random breath testing should be allowed and any driver found to have alcohol in his or her blood should be banned from driving for life.

8 Design and carry out a survey of car tyre tread conditions in the school car park. How would you present your results?

9 Design a test which would be carried out in a school laboratory to show that a 'crumple zone' at the front of a car reduces the risk of injury to passengers in an accident.

10 How could you investigate the power output of an electric motor when it is lifting different loads?

Chapter 3

1 Design and carry out an experiment to compare the strength of various cements made with different proportions of sand.
Find out what proportions of sand to cement are used by bricklayers. Are they always the same?

2 Design and carry out an investigation into the effect of acid rain on various building materials. Why do you think the cement around chimney stacks wears away quicker than on other parts of a building?

3 Design an experiment to show that moving water carries small rock particles more easily than large ones.

4 Design and carry out an investigation to compare slate and clay roofing tiles.

5 Design and carry out an investigation to test the effect on the elasticity of rock of the following:
   i)   temperature
   ii) force applied
   iii) length of time the force is applied.

6 'A joint or crack is a weak point.' How would you test this statement?
Why do cracks in roads get bigger?

7 Salt is said to have an effect on the strength of cement. How would you test this idea?

Chapter 4

1 Design an experiment to measure fuel consumption for a car under different driving conditions. If possible carry out this experiment for the family car (don't drive it yourself!) and compare the result with the manufacturer's value.

2 Draw a map of the world and then use an atlas to mark the sites of large coal deposits. What do you notice about the distribution of coal? What does this tell you about the Earth 500 000 000 years ago? How does coal affect a country's industry?

3 Design an experiment to compare the amount of ash left when different solid fuels are burnt in air.

4 If you have the opportunity to visit the Science Museum in London, make a study of how the introduction of coal gas changed the lives of people in Britain.

5 Carry out a survey to compare the costs of the various fuels made from crude oil. Include diesel (DERV), petrol, (leaded and unleaded), paraffin, and fuel oil.
Try to find out how much diesel costs for 'off-road' vehicles such as farm tractors etc. Suggest why this is.
Try to find out what the 'octane rating' for petrol means.

6 Use a gas bill to find out how much 1 MJ of energy costs when the fuel is natural gas. Compare this with the cost of 1 MJ of electricity. (Note: without guidance this may be awkward as the gas industry still gives volumes in cubic feet and still uses energy values quoted in BTU's. However, all the conversion factors are given on the bill . . . somewhere!)

7 Carry out a survey of old stone buildings in your town. Look for evidence of damage caused by smoke and/or acid attack. Make a list of buildings that have been cleaned by sand blasting . . . and ones that need it!

8 Design an experiment to measure the energy released by burning 'barbeque' charcoal.

9 Design an experiment to find the most cost-effective insulating material for use around a hot water tank. Your choice of materials might include polystyrene, mineral wool, or similar.

10 Try to find out what energy sources were used in your area in the past. Look for evidence of windmills, windpumps, watermills, etc.
If you can visit a windmill or a watermill find out about the technology used. For example, how was the speed controlled? How was the rotation of the sails or wheel used to drive machinery?

Chapter 5

1 Design an experiment to find how the beta-absorbing power of aluminium varies with its thickness.

2 Gamma radiation spreads out from its source just like light spreads out from a light bulb. Design an experiment to show how the amount of gamma radiation detected varies with the distance of the detector from the source.

3 Sketch a map of the United Kingdom. Mark on it the sites of the various nuclear power stations. Do you notice anything about where they are placed?

4 Design a questionnaire to find out what students in your school feel about the possession of nuclear weapons by countries.

5 Research the Three Mile Island disaster or the Chernobyl accident.

6 Find out about the following things and what radicarbon dating told us about them:
   a) Piltdown Man
   b) the Dead Sea Scrolls
   c) the Turin Shroud.

Chapter 6

1 'Adding sand to soil improves drainage'. Design and carry out an investigation to test this hypothesis. How might your results be of use to someone with clay soil in their garden?

2 Design and carry out an experiment to compare natural and artificial fertilizers. What are the advantages and disadvantages of each?

3 Design an experiment to investigate how much water an animal, e.g. a rabbit, drinks in a day., What things might affect this water intake?

4 Jack says that plants with differing leaf colours cannot photosynthesize. Jill disagrees. Who is right?

5 'Leaf size has no effect on the rate of photosynthesis'. How would you test this hypothesis.

6 How would you prove to someone that plants effectively 'breath out' carbon dioxide?

7 Design an experiment to show that the atmosphere contains various forms of pollution.

8 Does your town recycle waste materials? Carry out a survey to find out the answer to this question.

Chapter 7

1 Design and carry out an experiment to show how the resistance of a wire varies with its length.

2 Design and carry out an experiment to show how the resistance of a wire varies with its cross-sectional area.

3 Two battery manufacturers make batteries for use in bicycle lamps and torches. Each manufacturer claims that their batteries last longer and are better value for money. How could you independently test these claims?

4 A $10 \, \Omega$, $1 \, W$ carbon resistor is manufactured to carry a maximum curent of about $\frac{1}{3}$ A. Design an experiment to find how the resistance of the resistor varies as the current through it increases from $0 \, A$ to about $1 \, A$.

5 Investigate the cables used for wiring domestic lighting circuits, ring mains, and cooker circuits. comment on the materials used, the size of the wires, the cost per metre, etc.

6 Your parents say that the electricity bills are too high. Design and carry out a detailed study of where electricity is actually used in your house. Use it to check the electricity bill and to suggest how future bills might be reduced.

7 Carry out a consumer survey of mains plug design. Comment on the materials used, how the lid is fitted, how the cord is gripped, the design of the earth pin, and the cost of the plug.

8 An 'expert' says that clear lightbulbs give out more light and less heat than 'pearl' bulbs. Design an experiment to test this claim.

Chapter 8

1 Investigate whether the speed of a record deck's turntable varies with the load placed upon it.

2 Design and carry out an experiment to show how the resistance of a thermistor varies as it is heated from $0°C$ to $100°C$.

3 Design an experiment to find out whether radiator thermostatic valves can control room temperature accurately.

4 Find out how hypothermia affects particularly the elderly and mountain walkers. List the ways in which old people can be protected in cold weather. List the precautions that mountain walkers should take before setting out in poor weather.

5 A friend says that he knows all about feedback, and that it is a nuisance because he has heard the loud screeching noice made by an audio amplifier when a microphone is too near to it. Write down what you would say to him to explain the difference between negative and positive feedback and that negative feedback is very important to us.

6 Design an experiment to show how the resistance of an LDR varies with the brightness of the light falling on it.

Chapter 9

1 Design and carry out an experiment to show that living organisms release:
   a) heat energy
   b) carbon dioxide
      during respiration.

2 Design and carry out an experiment to see if yeast is present on the surface of fruit, e.g. grapes. Suggest why home-made jam sometimes smells of yeast.

3 How would you show that the bubbles in drinks like beer and champagne contain carbon dioxide?

4 Design and carry out an experiment to show that barley seeds contain starch and that this starch is converted to sugar during germination.

5 How would you extract the sugar from sugar cane or sugar beet?

6 Find out why wine sometimes turns to vinegar. Design and carry out an experiment to test this.

7 Design and carry out an experiment to test the hypothesis: 'pH has a direct effect upon the rate at which freshly sliced apples will go brown'. Why do apple slices dipped in lemon juice not go brown?

Chapter 10

1 Carry out a scientific comparison of a cheap washing up liquid and an expensive brand.

2 Shaving soap is designed to give a longer-lasting and creamier lather than ordinary washing soap. Design an experiment to find out whether this is true.

3 Many household cleaners are alkaline. Test the pH value of a selection of detergents, washing powders and floor cleaners. Compare your results with those for products designed for use on skin, i.e. soaps, shampoos, cleansing lotions, etc.

4 Find out about 'limestone pavements'. How were they formed and how are we destroying them?

5 Design a scientific test to find whether cold water washing powders are as good as hot water washing powders.

6 Washing powders for automatic washing machines are designed to give less lather than other powders. Design an experiment to find out whether this is true.

7 Washing-up liquid X washes more dishes for 1p than brand Y?

8 Biological powders remove egg stains better than non-biological powders?

Chapter 11

1 Use data from the weather reports printed in newspapers to compare the weather in the north, south, east and west of the United Kingdom. Display your results graphically and try to explain any general patterns you find.

2 Design an instrument which could be used to measure one of the following: cloud cover, hours of sunshine, wind speed.

3 Design an experiment to find whether a surface's texture is an important factor in the absorption or radiation of energy.

4 Design an experiment to show that the land (soil) heats up more quickly than the sea (water).

5 Find out about the effects of 'aerosol spray propellants' on the atmosphere. Design a poster warning people of the dangers and suggesting alternatives.

6 Find out about the effects of deforestation on the water cycle in places such as Bangladesh or Brazil.

7 Glass and polythene sheeting can be used as greenhouse 'glazing' materials. How could their 'effectiveness' be compared?

Chapter 12

1 What is the effect of using hormone rooting powder on root formation? Does the concentration make any difference?

2 Design an experiment to test the following ideas:
  a) 'brown hens tend to lay brown eggs'
  b) 'large hens tend to lay large eggs'
  c) 'people prefer brown eggs'
  d) 'brown eggs are more nutritious than white eggs'

3 Design an experiment to test the hypothesis: 'the genes for hair colour and eye colour are carried on the same chromosome'.

4 Design and carry out an investigation to see if insects are attracted more by the scent than by the colour of flowers.

5 Design an experiment to study the effect of temperature on the germination of seeds.

6 Design and carry out an experiment to test the following hypotheses:
  a) 'The seeds from some plant species need to be frozen before they will germinate'.
  b) 'The seeds from some plant species germinate better after they have been dipped in boiling water.